Theory and Method of Automatically Formed Roadway in Underground Mining

Qi Wang · Bei Jiang · Zhenhua Jiang

Theory and Method of Automatically Formed Roadway in Underground Mining

Springer

Qi Wang
China University of Mining
and Technology (Beijing)
Beijing, China

Bei Jiang
China University of Mining
and Technology (Beijing)
Beijing, China

Zhenhua Jiang
Shandong University
Jinan, China

ISBN 978-981-97-8081-5 ISBN 978-981-97-8082-2 (eBook)
https://doi.org/10.1007/978-981-97-8082-2

This Springer imprint is published by the registered company Springer Nature Singapore Pte Ltd.
The registered company address is: 152 Beach Road, #21-01/04 Gateway East, Singapore 189721, Singapore

Preface

The longwall mining technology is the primary application for underground mining of mineral resources, which offers benefits in ventilation effect and mining efficiency. The traditional longwall mining method involves excavating two mining roadways and leaving one mine pillar in one working face, which is called 121 mining method. The amount of roadway excavation is large, with approximately 13000 kilometers of roadway excavation per year in China, exceeding the diameter of the Earth. The stress is concentrated in where mine pillars are left, and mine pillar resources cannot be recovered. The 111 mining method was devised in response to the resource loss that resulted from the placement of mine pillars in the 121 mining method. One working face is mined using this approach, necessitating the excavation of one mining roadway in preparation. A filling material is used to form an artificial filling pillar instead of a mine pillar to maintain another roadway, reducing the amount of roadway excavation. However, the force transmission structure of the roadway roof remains unchanged, and there is still a problem of stress concentration in the artificial filling pillar.

In order to address the aforementioned issues, the balance mining theory is proposed in this book, which utilizes mining pressure and fragmentation-swelling characteristics of collapse gangue after underground mining. The traditional long cantilever beam structure is changed to a short cantilever beam structure, eliminating roadway excavation and mine pillar retention. Under the guidance of this theory, key technologies such as in-situ drilling and zoning technology, high prestressed energy absorption control technology, directional roof cutting technology, and roadway gangue rib support technology have been developed. The mining method of automatically formed roadway (AFR mining method) is successfully applied in a large number of mines with different buried depths, mine seam thicknesses, and roof types.

The layout of the book is as follows.

In Chap. 1, the definition and current status of the main longwall mining methods of underground mining were introduced. The current problems of this method in surrounding rock stress, excavation volume, and ecological protection were analyzed. The purpose and significance of developing AFR mining theory and method were clarified.

In Chap. 2, the gob-side entry driving method was analyzed in a non-balance state in terms of mine pillar stress, overlying rock structure movement, and roadway support system. The mining damage invariant equation was proposed, and a short cantilever beam structural model for mining was established. The broken expansion function equation and roadway broken expansion control equation for the roof collapse of AFR without mine pillars were obtained. Combining upper bound method, the mechanical model of AFR roof zoning control was established. It provided a theoretical basis for the design of AFR control parameters in different zones.

In Chap. 3, the corresponding technologies were developed according to balance mining theory. It mainly included in situ drilling and zoning technology, high prestressed energy absorption control technology, directional roof cutting technology, and roadway gangue rib support technology.

In Chap. 4, the comparative test study of numerical simulation between AFR mining method and traditional mining method was introduced. The distribution characteristics of stress in the mining area and the evolution law of roadway stress were analyzed. The influence of different parameters on the stability of AFR without mine pillars and the design basis for key parameters were elucidated.

In Chap. 5, the model test system of AFR without mine pillars with an integrated mining and roadway retention mode was developed. Geomechanical model tests were conducted on different types of roadways, such as shallow buried depth, deep buried depth, and fault structural areas. The control mechanism of AFR under different condition parameters was clarified.

In Chap. 6, the successful application of AFR mining method in different buried depths and roof types in mines was introduced. It effectively ensures the safety of mining, eliminates mine pillars retention, and reduces roadway excavation, which had certain safety and economic advantages.

In Chap. 7, the research hotspots and application scope of the theory, technology, and methods of AFR without mine pillars in recent years were introduced, and the future development of AFR mining method was prospected.

Beijing, China Qi Wang
Beijing, China Bei Jiang
Jinan, China Zhenhua Jiang

Acknowledgements This book has received funding from the National Key Research and Development Program of China (Grant No. 2023YFC2907600), the National Natural Science Foundation of China (Grant Nos. 52074164, 42477166, 42077267, and 42277174), and the Fundamental Research Funds for the Central Universities, China (Grant No. 2024JCCXSB01).

Contents

Chapter 1
Introduction

1.1 Background and Significance

The traditional and widely used longwall mining method is gob-side entry driving method (Chien 1981; Qian and Li 1982; Song et al. 1985). The method of mines involves mining one working face, requiring the advance excavation of two mining roadways for mine transportation and ventilation, and leaving one protective mine pillar to maintain the stability of the next mining roadway. So, the method is called 121 mining method, which was originated in England in 1706. The specific details of this method are shown in Fig. 1.1.

At present, the 121 mining method is widely used in underground mining, which has some characteristics. (i) The amount of roadway excavation is large. According to statistics, China's roadway excavation is about 13,000 km per year. The majority of the excavation process is manually completed, which limits the development of automated and intelligent mines. (ii) Collapse of the gob roof and the rotation & sinking of the rock beam cause stress concentration at the mine pillar, which can easily lead to deformation and damage of the surrounding rock of the roadway, and in severe cases, can cause dynamic disasters. (iii) During the mining process, there will be enormous mining pressure. Mining roadways primarily employ high strength support to withstand mining pressure, which can quickly lead to support component breakdown, leading to accidents such as roof collapse. (iv) Mining forms a large area of gob, and due to the support effect of mine pillars, the fracture and collapse forms of the gob and the overlying rock strata on the mine pillars are inconsistent. The continuous expansion of cracks to the surface can easily lead to uneven subsidence and environmental damage.

In 1937, Soviet scholars put forward the longwall mining gob-side entry retaining in view of the resource waste caused by 121 mining method. This method (Bai et al. 2004; Tan et al. 2015) involves mining one working face, requiring the advance excavation of one mining roadway, and the use of filling materials to form one artificial pillar instead of mine pillars to maintain another roadway as shown in Figs. 1.1 and

Q. Wang et al., *Theory and Method of Automatically Formed Roadway in Underground Mining*, https://doi.org/10.1007/978-981-97-8082-2_1

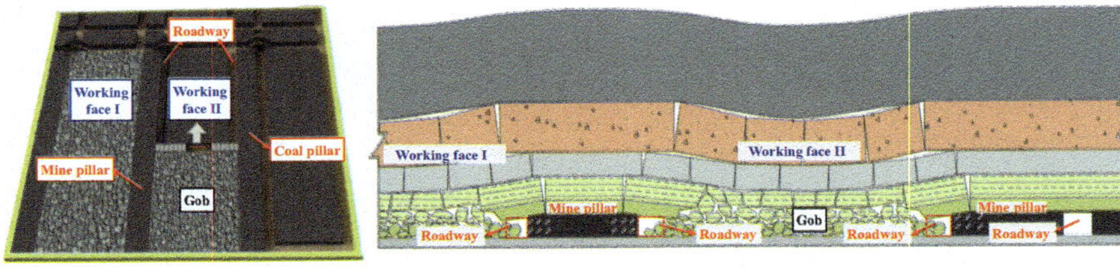

(a) Top view of working face (b) Section diagram of working face

Fig. 1.1 Schematic diagram of 121 mining method

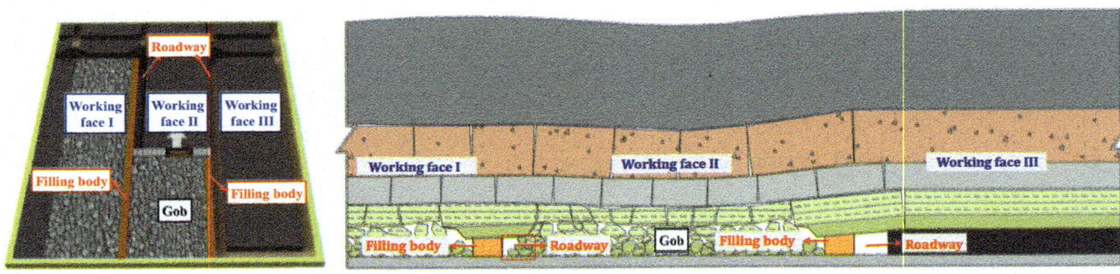

(a) Top view of working face (b) Section diagram of working face

Fig. 1.2 Schematic diagram of 111 mining method

1.2. So, the method is called 111 mining method. The 111 mining method has eliminated the retention of mine pillars, further improving the mine extraction rate and reducing the excavation of mining roadways (Zhang et al. 2012; Xie et al. 2020). However, this method still faces problems such as stress concentration in the filling body and failure of support components caused by huge mining pressure, which can easily lead to disasters such as surrounding rock large deformation, and restrict the large-scale promotion and application of the 111 mining method.

Many scholars have conducted in-depth research on the problems existing in the above two types of mining methods, focusing on roadway excavation, mine pillar and filling control, roadway support, and mining ecological protection. A series of theories and corresponding key technologies have been proposed, including loosening zone theory, stress control theory, combined support theory, strength weakening and shock-absorbing theory. However, various theories and technologies are mainly guided by the concept of using high-strength support to resist mining pressure, focusing on solving a certain aspect of the problems under the 121 and 111 mining methods systems. No unified study about balancing various problems is not carried out. There is a lack of transformative innovation in mining theory, resulting in difficulties in addressing the mining problems mentioned above. Therefore, the fundamental way to comprehensively solve problem of the mining process lies in transformation of mining methods.

After long-term exploration of mining rock mechanics, the co-author team developed a short cantilever beam (He et al. 2021) model for mining based on the idea of balance mining. The model combines the concepts of using rock mass broken expansion characteristics and turning resistance to mining pressure into mining pressure application. On this basis, an automatically formed roadway mining method (AFR mining method) is proposed. It utilizes mine pressure and broken expansion characteristics of collapsed gangue in underground space to eliminate reserved mine pillars and roadway excavation (Wang et al. 2018, 2020a, b, 2023). This method is different from traditional mining method which lies in transforming using high-strength support to resist mine pressure into using mine pressure to automatically form roadway. It solves the problem of stress concentration on the side of mine pillars from the source, and at the same time, this method belongs to balance mining, which can alleviate ground surface subsidence and protect the ecological environment.

Through continuous research and practice, the AFR mining method has achieved automatic roadway formation and safety control under different geological conditions. It has been successfully applied in various complex conditions such as different buried depths, roof types and so on. At present, it has been successfully applied in over 100 mines. In addition, it has achieved good application results and provided a safe & economical mining method, which can provide reference and theoretical guidance for similar projects.

1.2 The Main Content of This Book

This book mainly discusses the theory, key technologies, simulated tests, and on-site applications of AFR mining methods. The main research content is as follows.

(1) **The AFR mining theory**

To clarify the fundamental reasons for the non-balance state of traditional mining method, an analysis is conducted on the rock damage generated in each stage of the mining process from the perspective of mining rock mechanics. The mining damage invariant equation is proposed and solved, providing a foundation for balance control in mining activities. Based on balance mining concept, a short cantilever beam structural model for mining is established. The broken expansion function equation and roadway broken expansion control equation for the roof collapse of AFR without mine pillars are obtained. Combining upper bound method, the AFR surrounding rock zoning control concept is revealed. This provides a theoretical basis for safe mining without mine pillars in mines.

(2) **The AFR mining method and key technologies**

In order to achieve AFR without mine pillars, it is necessary to invention technologies according to balance mining theory. It mainly includes in-situ drilling and zoning technology, high prestressed energy absorption control technology, directional roof cutting technology, and roadway gangue rib support technology. The various technologies cooperate with each other to form AFR mining method without mine pillars. (a) This method utilizes in-situ drilling of surrounding rock to accurately obtain the mechanical parameters of the roadway surrounding rock, and guide roof cutting & roadway support parameters design. (b) Roadway stability during mining and completion stages is ensured by the high prestressed energy absorption control technology. (c) By directional cutting of the roof, the physical connection between the god roof and the roadway roof is quantitatively cut, and the mining pressure is used to cause the roof of the gob to collapse. The god is filled with broken expansion rock mass, forming a stable gangue rib. (d) Maintain the AFR by the roadway gangue rib support technology. Finally, this method eliminated the retention of mine pillars and automatically formed a roadway, while retaining the AFR for continued use in the next mining stage.

(3) **The control mechanism of the AFR mining method**

In order to clarify control mechanism of the AFR mining method, conduct a series of studies using numerical simulation and model testing methods. To meet the requirements of this research, a model test system of AFR without mine pillars with an integrated mining and roadway retention mode is developed. The numerical and model test research objects cover different types of engineering, which adopt AFR mining method and 121 mining method. The different types of engineering conditions include different buried depths, geological structures, and roof types. The AFR control mechanism with different conditions is clarified, which provides effective support for on-site design applications.

(4) **On site application of the AFR mining method**

The successful application of AFR mining method in various complex conditions such as different buried depths, and roof types in mines is introduced. At present, it has been successfully applied in over 100 mines. This book focuses on different types of mine working faces including shallow buried depth, deep buried depth, weathered roof strata, broken roof strata, near the overlying mined coal seam and other typical examples. It effectively ensures the coal mining safety, eliminates mine pillars, and reduces roadway excavation, which has certain safety and economic advantages.

References

Bai JB, Zhou HQ, Hou CJ, Tu XZ, Yue DZ (2004) Development of support technology beside roadway in goaf-side entry retaining for next sublevel. J China Univ Min Technol 33(2):59–62 (in Chinese)

Chien MG (1981) A study of the behaviour of overlying strata in longwall mining and its application to strata control. Dev Geotech Eng 32:13–17

He MC, Wang Q, Wu QY (2021) Innovation and future of mining rock mechanics. J Rock Mech Geotech Eng 13(1):1–21

Qian MG, Li HC (1982) Conditions required for equilibrium of overlying strata at working areas. J China Coal Soc 2:1–12 (in Chinese)

Song ZQ, Jiang JQ, Song Y (1985) Design of strata control in roadways ground working faces. J Shandong Univ Sci Technol (Nat Sci) 3:1–11 (in Chinese)

Tan YL, Yu FH, Ning JG, Zhao TB (2015) Design and construction of entry retaining wall along a gob side under hard roof strata. Int J Rock Mech Min Sci 77:115–121

Wang Q, He MC, Yang J, Gao HK, Jiang B, Yu HC (2018) Study of a no-pillar mining technology with automatically formed gob-side entry retaining for longwall mining in coal mines. Int J Rock Mech Min Sci 110:1–8

Wang Q, Zhang P, Jiang ZH, He MC, Li SC, Wang Y, Jiang B (2020a) Automatic roadway formation method by roof cutting with high strength bolt-grouting in deep coal mine and its validation. J China Coal Soc 46(2):382–397 (in Chinese)

Wang Q, Jiang B, Xin ZX, He MC, Li SC, Zhang P, Wang Y, Jiang ZH (2020b) Development of a 3D geomechanical model test system for non-pillar mining with automatically formed roadway and its engineering application. Chin J Rock Mech Eng 39(8):1582–1594 (in Chinese)

Wang Q, Jiang ZH, Jiang B et al (2023) Ground control method of using roof cutting pressure release and energy-absorbing reinforcement for roadway with extra-thick hard roof. Rock Mech Rock Eng 56(10):7197–7215

Xie SR, Pan H, Chen DD, Zeng JC, Hai ZS, Cheng Q, Xiao HB, Yan ZQ, Li YH (2020) Stability analysis of integral load-bearing structure of surrounding rock of gob-side entry retention with flexible concrete formwork. Tunn Undergr Space Technol 103:103492

Zhang N, Yuan L, Han CL, Xue JH, Kan JG (2012) Stability and deformation of surrounding rock in pillarless gob-side entry retaining. Saf Sci 50(4):593–599

Chapter 2
Mining Theory of Automatically Formed Roadway (AFR)

Through in-depth analysis of the mine pressure behaviors law at different stages of the traditional 121 mining method, the stress characteristics of the long cantilever beam structure of this method are clarified. Meanwhile, a fundamental solution to the problems of traditional mining technology is sought from the perspective of rock mechanics.

2.1 The Characteristics of the Traditional Mining Method

2.1.1 Mine Pressure Behaviors at Different Stages

Traditional longwall 121 mining method can be divided into five stages (Stage I–V) during mining process in entire working face.

Stage I: The immediate roof is an unstable rock strata above the mine seam. During the mining process, the stress in the surrounding rock of the gob roof redistributes, but immediate roof does not collapse immediately, and the rock above it does not deform, as shown in Fig. 2.1b.

Stage II: After mining at a certain distance behind the working face, immediate roof collapse occurs under its own weight & mining stress influence, and impact generated by collapse has a certain impact on the stability of the roadway. At this time, the gangue formed by immediate roof collapse failed to fill the gob, and the rock strata above it underwent slight deformation without obvious cracks, as shown in Fig. 2.1c.

Stage III: The basic roof is a stable rock strata located above immediate roof and capable of maintaining a large overhanging area. Due to the fact that the gob is not filled with collapsed gangue, the basic roof fractures and rotates to form a caving zone above the gob. Obvious cracks appear inside and begin to develop towards the rock strata above it, as shown in Fig. 2.1d. At this point, under the action of

Q. Wang et al., *Theory and Method of Automatically Formed Roadway in Underground Mining*, https://doi.org/10.1007/978-981-97-8082-2_2

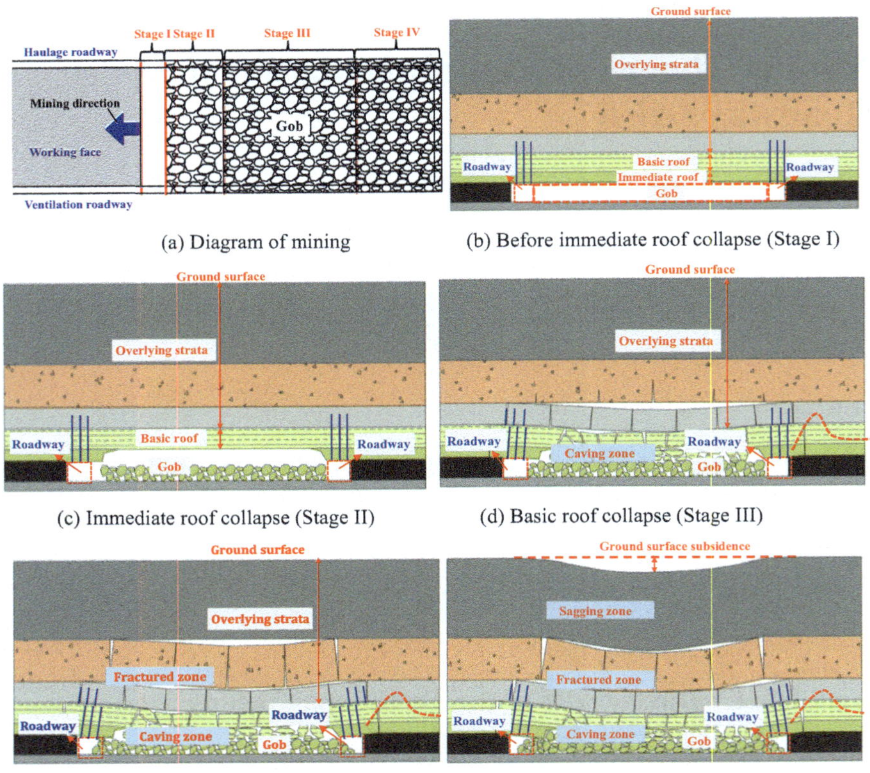

(a) Diagram of mining (b) Before immediate roof collapse (Stage I)

(c) Immediate roof collapse (Stage II) (d) Basic roof collapse (Stage III)

(e) Overlying strata fracture and subsidence (Stage IV) (f) Ground surface subsidence (Stage V)

Fig. 2.1 Schematic diagram of mining in each stage

basic roof rotation compression, the stress of the surrounding rock in the roadway significantly increases, and a stress concentration area appears surrounding rock around the roadway.

Stage IV: As the working face mining, due to the periodic collapse of the basic roof along the mining direction of the working face, the rock strata above it fractures and sinks, and the cracks rapidly develop, forming a fractured zone. The stress peak is progressively moving towards the deeper surrounding rock of the roadway, but the stress concentration region on the mine side of the roadway is still present, as shown in Fig. 2.1e.

Stage V: After the mining of the working face is completed, the rock strata above the fracture zone shows bending and sinking, forming a sagging zone. As time goes by, it leads to surface subsidence and crack damage, as shown in Fig. 2.1f. At this point, the rock mass of this working face is temporarily stable.

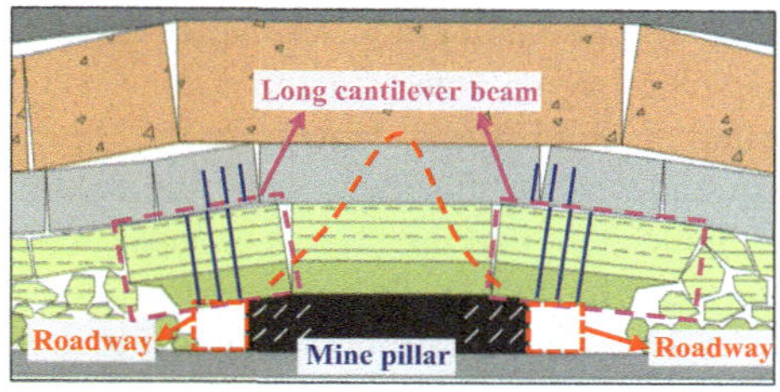

Fig. 2.2 Long cantilever beam structure in 121 mining method

2.1.2 The Long Cantilever Beam Mining Model

After going through the above five stages, the working face formed a long cantilever beam roof structure at the roadway. The specific structure of the long cantilever beam is as shown in Fig. 2.2.

(1) In terms of the stress characteristics of mine pillars: During the mining of adjacent working faces, the overlying rock strata in the gob experience stage collapse and fracture sinking, forming a long cantilever beam structure above the roadway on both sides of the mine pillar. The long cantilever beam undergoes rotation and sinking, squeezing the mine pillar in the roadway area, resulting in stress concentration in the mine pillar area, making it difficult to release stress. Over time, the mine pillar is prone to gradually being crushed.
(2) In terms of the movement law of the overlying rock structure: After the mining of the working face, the rock strata above the mine pillar gradually produces tensile failure due to the support effect of the mine pillar. At the same time, the fracture and collapse morphology of the gob and the overlying rock strata on the pillar are inconsistent. Overlying rock structure is constantly moving. So that the overall stability time is very long, and the overlying rocks are always in non-balance state.
(3) In terms of the force state of the roadway support system: To ensure the safety of the mining roadway, high-strength support is used in this method to withstand mining stress. Under enormous mining pressure, the roadway support system is subjected to non-balance forces, which can easily lead to roadway control problems.

2.2 The AFR Mining Theory

2.2.1 The Invariant Equation for Mining Damage

In order to identify the cause of the aforementioned non-balance state of the 121 mining technique, mining rock mechanics is employed to examine the rock strata damage at each stage of mining activity. Mining activities frequently causes large area collapse and crack formation in the rock strata above the gob. The roof strata develop caving zone, fractured zone, and sagging zone that causes ground surface sinking. Rock strata damage in the three zones creates continuous changes in the ground surface subsidence variable volume $\triangle V_S$, the roof strata fracture expanding volume $\triangle V_C$ and the broken expansion volume of the collapsed rock mass $\triangle V_B$, as seen in Fig. 2.3.

When the rock strata damage (Fig. 2.3) and 121 mining technique (Fig. 2.1) are combined, the mining process at each stage is described as follows.

(1) Stage I (before immediate roof collapse): $\triangle V_B = \triangle V_C = \triangle V_S = 0$.
(2) Stage II (immediate roof collapse): $\triangle V_B < \triangle V_m$, $\triangle V_C = 0$.
(3) Stage III (basic roof collapse): $\triangle V_B$ keeps rising. At this point, $\triangle V_C$ starts to rise and $\triangle V_B < \triangle V_m$.
(4) Stage IV (overlying strata fracture and subsidence): $\triangle V_B$ stays constant while $\triangle V_C$ rises quickly.
(5) Stage V (ground surface subsidence): $\triangle V_S$ starts to rise.

To characterize the damage status of various rock strata, three categories of damage variables are established. Equation (2.1) states that K_1 is the damage variable for ground surface sinking brought on by mining. It is computed as the ratio of the mining volume to the volume of the ground surface subsidence variable. Equation (2.2) states that K_2, the ratio of the mining volume to the increasing fracture volume of the roof strata, is the damage variable of roof strata fracture. Ratio of collapsed rock mass's broken expansion volume to its mining volume, or K_3, is rock mass's broken expansion damage variable Eq. (2.3).

$$K_1 = \Delta V_S / \Delta V_m \tag{2.1}$$

$$K_2 = \Delta V_C / \Delta V_m \tag{2.2}$$

$$K_3 = \Delta V_B / \Delta V_m \tag{2.3}$$

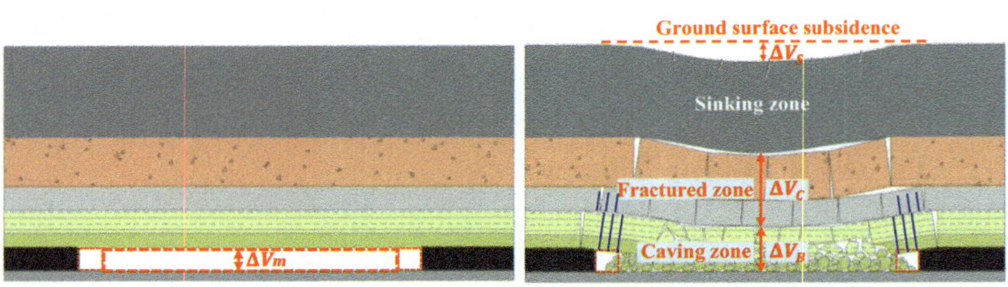

(a) Before roof collapse (b) After roof collapse

Fig. 2.3 The damage diagram of the rock strata

The three categories of rock strata damage variables mentioned above are subject to continual change depending on specific geological circumstances. Despite the complexity of the three damage variables' variations, their total always adds up to one. The basis for all mining engineering methods is this equation, which is referred to as the invariant equation of the sum of the mining damage variables. It is known as mining damage invariant equation (He et al. 2021).

$$K_1 + K_2 + K_3 = 1 \tag{2.4}$$

The mining volume $\triangle V_m$ of the 121 mining technique is known and may be found in Eqs. (2.1)–(2.4). Measurements may be used to find the value of K_1 and produce ground subsidence variable volume $\triangle V_S$. However, related mining damage variables K_2 and K_3 cannot be identified because the roof strata fracture growing volume $\triangle V_C$ and the broken expansion volume $\triangle V_B$ are not controlled, as shown in Eq. (2.4).

The fractured rock mass displays broken expansion characteristic. In principle, $\triangle V_B$ can equal $\triangle V_m$ ($\triangle V_B = \triangle V_m$) if the features can be employed scientifically to realize balance between the broken expansion volume of the collapsed rock mass and the mining volume. The ground surface sinking damage variable K_1 and fracture damage variable K_2 are computed as 0 ($K_1 = K_2 = 0$) and the broken expansion damage variable K_3 is calculated as 1 ($K_3 = 1$) by putting this value into Eqs. (2.3) and (2.4). By solving the mining damage invariant equation Eq. (2.4), one may regulate the balance of mining operations and create a balance mining system.

2.2.2 *The Short Cantilever Beam Mining Model*

Based on previously discussed idea of balance mining, a short cantilever beam model was developed, as seen in Fig. 2.4. The concept of using the fractured expansion properties of the fractured rock mass is combined with the idea of using mining pressure to overcome resistance to it. On the one hand, directed splitting on gob-side roof allows mining pressure to cause roof directional collapse towards gob, and the gangue's broken expansion properties are used to fill the gob with gangue. Because of this, the gangue rib develops spontaneously in the subterranean area that is exposed during mining. However, to preserve gangue rib, gangue blocking support is utilized. Roadway's roof is controlled by energy absorption support, which guarantees stability of the road around the rocks. It is evident that the collapsed rock mass's broken expansion characteristics and mining pressure are used, and that the mine pillar reserve and road excavation are eliminated. The balance mining theory with an integrated mode of mining and roadway retaining is established based on the short cantilever beam model.

The balance mining concept can be used to fill the gob with the broken expansion of gangue and cause the working face's overlying strata to collapse directionally. As a result, it is possible to achieve both the balance force of roadway support system and the quick balance of the rock strata structure above. According to the function equation of broken expansion (Eq. 2.6) obtained from field measurements and the

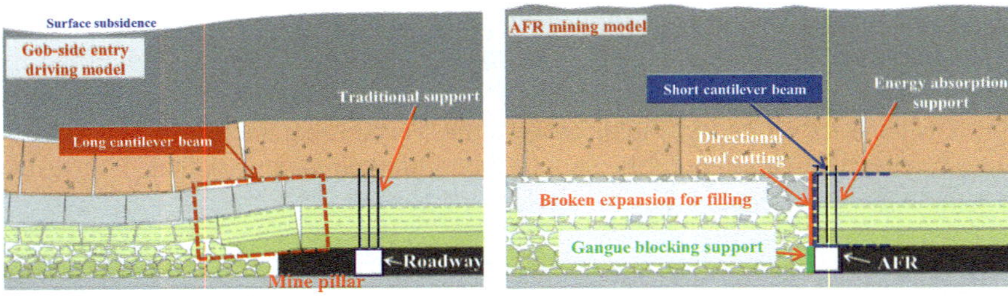

Fig. 2.4 The long and short cantilever beam mining models

roadway broken expansion control equation of the collapsed rock mass (Eq. 2.5), an appropriate roof cutting height is meant to control the broken expansion volume of the collapsed rock mass. When $V_B = \triangle V_m$, $\triangle V_C = 0$, and $\triangle V_S = 0$, it achieves balance between the broken expansion volume and mining volume. The aforesaid mining method, as seen in Fig. 2.5, may realize overall balance of each roof cutting region.

Broken expansion control equation Eq. (2.5) is used for the calculation of broken expansion volume $\triangle V_B$ of the collapsed rock mass in the AFR.

$$\Delta V_B = (K - 1)H_C * S \tag{2.5}$$

where S is the mining area, H_C is the height of roof cutting, and K is the broken expansion coefficient of the collapsing rock mass.

Function equation of broken expansion Eq. (2.6) is used for the calculation of the parameter K.

$$K = K_0 e^{-\alpha t} \tag{2.6}$$

where t is the time, α is fitting coefficient. K_0 is initial broken expansion coefficient.

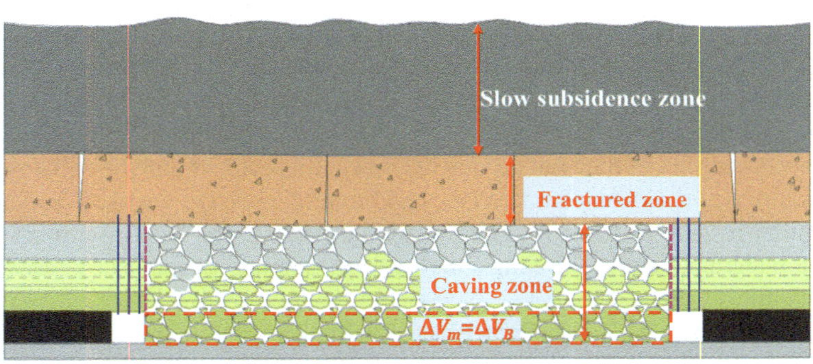

Fig. 2.5 Mining damage invariant model (He et al. 2021)

2.2.3 The AFR Balance Mining Characteristics

Figure 2.6 depicts the features of the AFR's balance mining system without pillars. After mining, the working face's balance state primarily takes three forms: the gob automatic balance, the roadway balance by using artificial assistance, the overall balance of several working faces. These features are explained below.

(1) **Gob automatic balance**

The broken expansion balance during mining is shown by the gob's overlaying strata. This balance state is an automated one that is created by using the collapsed rock mass's broken expansion features and mining pressure. The gob-side roof undergoes directional splitting to allow mining pressure & deadweight of underlying strata to cause them to collapse within the roof cutting height. Continuous strata collapse and broken gangue expansion fill the gob, guaranteeing that the amount of broken expansion equals the volume of mining. As a result, gob automatic balance is achievable.

(2) **Roadway balance by using artificial assistance**

Here is the particular performance: (a) To lessen the impact from gob roof movement and collapse, short cantilever beam construction is produced by roof cutting. (b) According to rock broken expansion characteristics, collapsed rock mass in the gob contributes to support roadway short cantilever beam roof in a specific way. (c) The roadway's gangue rib receives gangue blocking maintenance, and high prestressed energy absorbing anchor cable is used to support the roadway roof. As a result, artificial intervention achieves roadway balance.

(3) **Overall balance of several working faces**

After mining, the numerous working faces exhibit general balance, preventing the effects of roadway excavation and mine pillar reserve on working faces. Here is the particular performance: (a) There is no longer a mine pillar reserve. Uneven ground surface sinking is prevented when fracturing and collapse occur in the rock strata

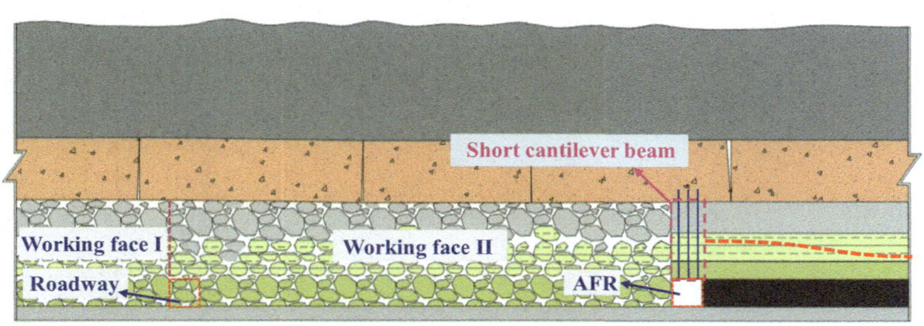

Fig. 2.6 Mining diagram of the AFR without pillar

because this eliminates the support of mine pillars. (b) There is no need to excavate the roadway. Overall balance time of mining area can be significantly decreased by avoiding the impact of roadway excavation in the neighboring working face on balance region of prior working face. As a result, it is possible to achieve the overall balance of several working faces.

2.2.4 The AFR Zoning Control Concept

In addition to stopping the stress transfer between the roadway roof and the gob, the AFR technique without mine pillars also converts the roadway roof's rock structure into a short cantilever beam structure. The key to successfully implementing the AFR approach is the stability of the short cantilever beam construction. This part proposed the mechanical model of the AFR roof zoning control and describes the roof fracture mechanism using upper limit method. It provides a theoretical basis for the design of AFR control parameters in different partitions.

(1) **Evolution pattern of the roof structure of the AFR**

According to the structural characteristics of the surrounding rock of the roof and gob of the AFR in the process of advancing the working face, and combined with the mechanical effects of a pre-splitting on the rock mass of the roof, the roadway is divided into three zones: the pre-splitting implementation zone, the pre-splitting influence zone, and the pre-splitting stabilisation zone, as shown in Fig. 2.7.

1) Pre-splitting implementation zone (Zone I): After pre-splitting is carried out on the roof of the AFR, the pre-splitting runs through the immediate roof, cutting

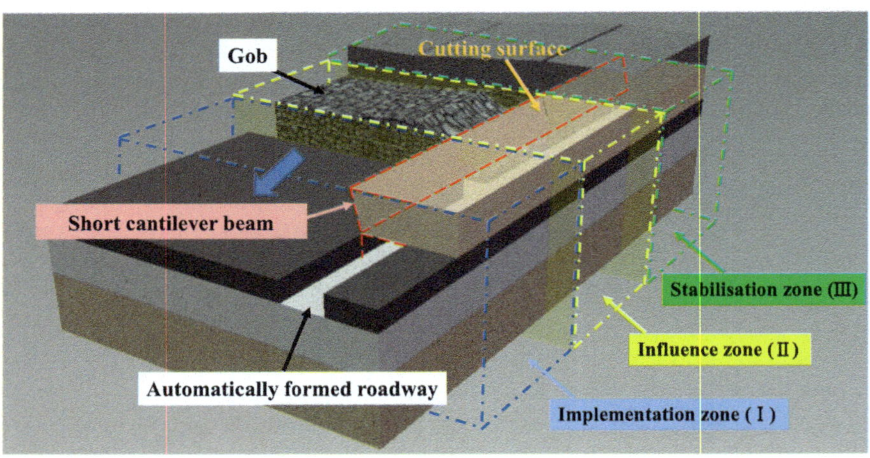

Fig. 2.7 Roof structure characteristics of the AFR. Reprinted from Ref. Wang et al. (2020), Copyright © 2020, Saudi Society for Geosciences

off the connection between the roof of the roadway and the immediate roof of the working face. Consequently, the roof of the AFR forms a short cantilever beam structure. However, the working face has not yet been mined in this zone, so the immediate roof of the working face provides a certain support to the roof of the roadway, ensuring that there is no large deformation of the roadway roof in the presence of the pre-splitting.

2) Pre-splitting influence zone (Zone II): After the working face has been mined past the pre-splitting, the immediate roof falls along the pre-splitting, causing the basic roof to be in the suspended and exposed state. When the rock mass of the basic roof reaches the ultimate stress, the basic roof fractures, bends, and sags. In this zone, the falling, bending and sagging of the roof do not stabilise, and the falling gangue of the gob is not entirely in contact with the short cantilever beam, so there is a certain amount of unfilled space.
3) Pre-splitting stabilisation zone (Zone III): As the working face is mined, the basic roof continues to bend and sag slowly and finally touches the gangue. The falling gangue of the gob is in complete contact with the short cantilever beam, and the falling gangue is compacted, achieving an equilibrium state.

(2) **Zoning control model of short cantilever beam**

As seen in Fig. 2.8, fracture mechanics model of AFR is established. The short cantilever beam is h in height and $2l$ in width, and the fracture curve is $f(x)$. The ensuing presumptions are made: The surrounding rock stress, denoted by q, is distributed evenly along the curve $f(x)$ and is perpendicular to the fractured roof. The pre-tightening force of a single anchor cable of the roadway roof is p_a, the gob-side roof support force is F, and the force of the solid roadway side to the roof is p_c.

Since the AFR is impacted when mining both the current and the next working face, it is imperative to examine how mining is impacting the fracture range of the

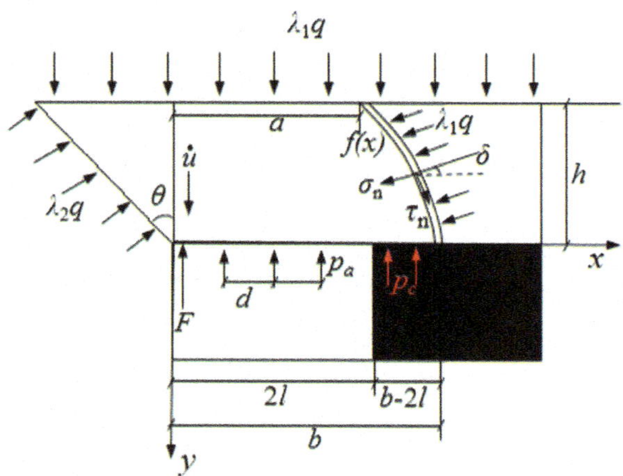

Fig. 2.8 Model of fracture mechanics for the AFR. Reprinted from Ref. Wang et al. (2020), Copyright © 2020, Saudi Society for Geosciences

short cantilever beam. The tension in the rock mass surrounding the fractured roof may be represented as λ_{1q}, where λ_1 is the mining impact coefficient and $\lambda_1 > 1$. The mining of the working face in the AFR generated three zones: the pre-splitting implementation zone, pre-splitting influence zone, and pre-splitting stabilization zone. Pre-splitting influence zone is the stage with the greatest impact on roadway safety. Support force of descending gangue varies continuously as the gangue rib it forms progressively makes touch with the immediate roof. The support force of the broken expand gangue is set to λ_{2q}, where λ_2 is the support coefficient of the broken expand gangue ($0 < \lambda_2 < 1$), and the direction of λ_{2q} is perpendicular to the pre-splitting plane, in order to represent the support impact of the gangue rib on the short cantilever beam.

According to the virtual power equation, when the external force work power is higher than the internal energy dissipation rate of the rock mass, the roof is fractured and fails. The kinematically admissible velocity field is established, the velocity direction is downward along the y-axis, and the velocity vector is $\dot{u}$.

If the fracture surface of the rock mass is regarded as a thin deformed layer with a certain thickness, the internal energy dissipation rate is only generated at this thin separation layer. According to the Hoek–Brown failure criterion and its associated flow rule, the plastic shear strain rate $\dot{\gamma}^p$ and the plastic positive strain rate $\dot{\varepsilon}^p$ at the fracture surface of the rock mass can be calculated, and then the energy dissipation rate per unit volume in the thin deformed layer can be obtained.

According to the Hoek–Brown criterion, the yield function F can be expressed as

$$G = \tau_n - A\sigma_c\left[(\sigma_n + \sigma_t)\sigma_c^{-1}\right]^B \tag{2.7}$$

According to the associated flow rule

$$\dot{\varepsilon}_{ij} = \dot{\lambda}\frac{\partial Q}{\partial \sigma_{ij}} \tag{2.8}$$

where Q is a plastic potential function, and $\dot{\lambda}$ is a plasticity factor.

Let $Q = G$. Substituting Eq. (2.7) into Eq. (2.8) gives the plastic strain rate on the thin separation layer:

$$\begin{cases} \dot{\varepsilon}_n = \dot{\lambda}\frac{\partial Q}{\partial \sigma} = -\dot{\lambda}AB\left[(\sigma_n + \sigma_t)\sigma_c^{-1}\right]^{B-1} \\ \dot{\gamma}_n = \dot{\lambda}\frac{\partial Q}{\partial \tau} = \dot{\lambda} \end{cases} \tag{2.9}$$

The formula for calculating the plastic strain rate is derived as

$$\begin{cases} \dot{\varepsilon}_n = \frac{\dot{u}}{w\sqrt{1+f'(x)^2}} \\ \dot{\gamma}_n = -\frac{\dot{u}f'(x)}{w\sqrt{1+f'(x)^2}} \end{cases} \tag{2.10}$$

Equations (2.9) and (2.10) are combined to get:

$$-\frac{\dot{u}}{\dot{\lambda}w\sqrt{1+f'(x)^2}}=AB\left[(\sigma_n+\sigma_t)\sigma_c^{-1}\right]^{B-1} \tag{2.11}$$

Equation (2.11) can be simplified to get the normal stress on the thin separation layer:

$$\sigma_n=\sigma_c\left[f'(x)AB\right]^{\frac{1}{1-B}}-\sigma_t \tag{2.12}$$

$$\dot{D}=\sigma_n\dot{\varepsilon}_n+\tau_n\dot{\gamma}_n=\left\{-\sigma t+\sigma c\left[ABf'(x)\right]^{\frac{1}{1-B}}\left(1-B^{-1}\right)\right\}\dot{u}/w\sqrt{1+(f'(x))^2} \tag{2.13}$$

It can be seen from Fig. 2.8 that in the interval [a, b], when the surrounding rock of the roadway roof fractures, the internal energy dissipation rate at the deformed thin layer of the fracture surface can be expressed as

$$\dot{W}_{\mathrm{D}}=\int_0^S\left\{\sigma t-\sigma c\left[f'(x)AB\right]^{\frac{1}{1-B}}\left(1-B^{-1}\right)\right\}\dot{u}/\sqrt{1+(f'(x))^2}ds \tag{2.14}$$

where S is the total arc length of the curve $f(x)$ in the interval [a, b] and satisfies $ds=\sqrt{1+f'(x)^2}dx$.

External force power calculation: The external force power is composed of six parts: the work done by gravity, the surrounding rock stress q, the anchor cable pre-tightening force p_a, the gob-side roof support force F, the gangue rib p_g, and the solid roadway side p_c.

Power of gravity $\dot{W}_\gamma$.

Let the rock unit weight be γ. The gravity work power is

$$\dot{W}_\gamma=-\int_a^b\gamma f(x)dx\cdot\dot{u}+\gamma ah\dot{u}+\frac{1}{2}h^2\tan\theta\dot{u} \tag{2.15}$$

Power of the surrounding rock stress $\dot{W}_q$

$$\dot{W}_q=\left(\int_0^S\lambda_1q\sin\delta ds+\lambda_1qh\tan\theta\right)\dot{u}=\lambda_1q(b+h\tan\theta)\dot{u} \tag{2.16}$$

Power of the pre-tightening force of the anchor cable $\dot{W}_{pa}$.

Let the spacing between the anchor cables be d. Then, within the fracture range of the short cantilever beam, the required number of anchor cables is

$$n=b/d+1 \tag{2.17}$$

If the result calculated by Eq. (2.17) is a decimal, n takes the integer part, and the power of the pre-tightening force of the anchor cable is

$$\dot{W}_{pa} = -(n - 2 + 2\cos\theta)p_a\dot{u} \tag{2.18}$$

Power of the gob-side roof support force $\dot{W}_F$.

When the roof of the roadway is cut to form a short cantilever beam, pillars are installed on the working face side of the roadway to support the roof. The power of the gob-side roof support force is

$$\dot{W}_F = -F\dot{u} \tag{2.19}$$

Power of the gangue rib $\dot{W}_g$

$$\dot{W}_g = \int_0^S \lambda_2 qh\tan\theta\dot{u} = \lambda_2 qh\tan\theta\dot{u} \tag{2.20}$$

Power of the solid roadway side $\dot{W}_{pc}$

$$\dot{W}_{pc} = -p_c(b - 2l)\dot{u} \tag{2.21}$$

where p_c is the compressive strength of the coal mass of the roadway side, and $b - 2l$ is the influence range of the fractured roof on the solid roadway side.

The total energy dissipation power of the short cantilever beam is

$$\begin{aligned}
\xi &= \dot{W}_D + \dot{W}_\gamma + \dot{W}_q + \dot{W}_{pa} + \dot{W}_F + \dot{W}_g + \dot{W}_{pc} \\
&= \int_a^b \left[\sigma_t - \sigma_c(AB)^{\frac{1}{1-B}}(1 - B^{-1})f'(x)^{\frac{1}{1-B}}\right]dx \cdot \dot{u}\left\{\int_a^b \gamma f(x)dx - \gamma ah - \frac{1}{2}h^2\tan\theta\right. \\
&\quad - \lambda_1 q(b + h\tan\theta) + (b/d - 1 + 2\cos\theta)p_a \\
&\quad + F + \lambda_2 qh\tan\theta + p_c(b - 2l)\} \cdot \dot{u}
\end{aligned} \tag{2.22}$$

That is,

$$\begin{aligned}
\xi = &\left\{\int_a^b \varphi\left[f'(x), f(x), x\right]dx - \gamma ah - \frac{1}{2}h^2\tan\theta - \lambda_1 q(b + h\tan\theta)\right. \\
&+ (b/d - 1 + 2\cos\theta)p_a + F + \lambda_2 qh\tan\theta + p_c(b - 2l)\} \cdot \dot{u}
\end{aligned} \tag{2.23}$$

According to the variational method, the equation of the fracture curve of the short cantilever beam is derived as

$$f(x) = \sigma_c^{B-1/B} A^{-1/B} \gamma^{-1} (\gamma x + c_1)^{1/B} + c_2 \tag{2.24}$$

where c_1 and c_2 are undetermined integral constants.

An infinitesimal element is taken at the junction of the short cantilever beam and the basic roof to conduct the mechanical analysis and establish the mechanical equilibrium equation of the infinitesimal element. Combined with the tangential stress and the normal stress of the velocity discontinuity line, the shear stress of the infinitesimal element τ_{xy} is obtained as

$$\tau_{xy} = \frac{1}{2}\sigma_n \sin(2\delta) - \tau_n \cos(2\delta) \tag{2.25}$$

When the short cantilever beam fractures, the shear stress between the beam and the basic roof is zero.

$$\tau_{xy}(x = a, y = -h) = 0 \tag{2.26}$$

$$\begin{aligned} &-\sigma_t \sigma_c^{(B-1)/B} A^{-1/B} B^{-1} (\gamma a + c_1)^{(1-B)/B} + \sigma_c^{2(B-1)/B} A^{-2/B} B^{-1} (\gamma a + c_1)^{(2-B)/B} \\ &- (\gamma a + c_1)\left[\sigma_c^{2(B-1)/B} A^{-2/B} B^{-2} (\gamma a + c_1)^{2(1-B)/B} - 1\right] = 0 \end{aligned} \tag{2.27}$$

To satisfy Eq. (2.27),

$$c_1 = -\gamma a \tag{2.28}$$

Substituting Eq. (2.28) into Eq. (2.24), we have

$$f(x) = \sigma_c^{(B-1)/B} A^{-1/B} \gamma^{(1-B)/B} (x - a)^{1/B} + c_2 \tag{2.29}$$

The boundary conditions of the failure curve in the interval [a, b] are

$$\begin{cases} f(x = a) = -h \\ f(x = b) = 0 \end{cases} \tag{2.30}$$

Substituting Eq. (2.29) into Eq. (2.30), we obtain

$$\begin{cases} c_2 = -h \\ \sigma_c^{(B-1)/B} A^{-1/B} \gamma^{(1-B)/B} (b - a)^{1/B} + c_2 = 0 \end{cases} \tag{2.31}$$

The fracture curve equation of the short cantilever beam conforming to the generalized Hoek–Brown criterion can be obtained as

$$f(x) = \sigma_c^{(B-1)/B} A^{-1/B} \gamma^{(1-B)/B} (x - a)^{1/B} - h \tag{2.32}$$

According to the energy conversion theory, let the internal energy dissipation rate be equal to the external force power, that is $\xi = 0$. Thus, combining (2.32) and (2.23), the equation for the support force along gob side of the AFR is as follows.

$$
\begin{aligned}
F = {} & \frac{B}{1+B}\sigma_c^{(B-1)/B} A^{-1/B} \gamma^{1/B} (b-a)^{(1+B)/B} + \gamma ah + \frac{1}{2}h^2 \tan\theta \\
& + \lambda_1 q(b + h\tan\theta) - \sigma_t(b-a) - \left(b/d - 1 + 2\cos\theta\right)p_a \\
& - \lambda_2 qh\tan\theta - p_c(b-2l)
\end{aligned}
\tag{2.33}
$$

(3) **Design basis for support force along gob side of the AFR in different zones**

According to the stress characteristics of the short cantilever beam roof at different zones of the AFR mentioned above and Eq. (2.33), support forces are designed for three zones including the pre-splitting implementation zone, the pre-splitting influence zone and the pre-splitting stabilisation zone.

1) **Support force design in the pre-splitting implementation zone**

When the AFR is in the pre-splitting implementation zone, the rock strata in the gob begin to collapse layer by layer, and the support impact of collapsed gangue for the AFR roof is relatively little. At this time, the rock mass is relatively loose, and the supporting effect of the collapsed rock mass can be ignored in the early stage of rock strata collapse. The supporting force of the collapsed rock mass for the AFR roof is not considered, that is $\lambda_2 = 0$. Considering the advanced influence of mining on the working face, the critical value of support force along gob side of the AFR in the pre-splitting implementation zone can be obtained. The specific design basis for support force along gob side of the AFR in the pre-splitting implementation zone is in Eq. (2.34).

$$
\begin{aligned}
F = {} & \frac{B}{1+B}\sigma_c^{(B-1)/B} A^{-1/B} \gamma^{1/B} (b-a)^{(1+B)/B} + \gamma ah + \frac{1}{2}h^2 \tan\theta \\
& + \lambda_1 q(b + h\tan\theta) - \sigma_t(b-a) - \left(b/d - 1 + 2\cos\theta\right)p_a - p_c(b-2l)
\end{aligned}
\tag{2.34}
$$

2) **Support force design in the pre-splitting influence zone**

As the working face advances, the basic roof rock mass reaches its ultimate stress and breaks. During the process of the basic roof breaking and sinking, mine pressure is applied to the lower rock strata and AFR, and the broken rock mass in the gob is continuously compacted. Considering the mining influence of the working face and supporting effect of broken expansive rock mass, the critical value of support force along gob side of the AFR in the pre-splitting influence zone can be obtained. The specific design basis for support force along gob side of the AFR in the pre-splitting influence zone is in Eq. (2.35).

$$F = \frac{B}{1+B}\sigma_c^{(B-1)/B}A^{-1/B}\gamma^{1/B}(b-a)^{(1+B)/B} + \gamma ah + \frac{1}{2}h^2\tan\theta$$
$$+ \lambda_1 q(b + h\tan\theta) - \sigma_t(b-a) - \left(b/d - 1 + 2\cos\theta\right)p_a$$
$$- \lambda_2 qh\tan\theta - p_c(b-2l) \quad (2.35)$$

3) **Support force design in the pre-splitting stabilisation zone**

As the length of the AFR reaches a certain distance, the collapsed rock mass in the gob are compacted and tend to stabilize. By utilizing the stable support force of the broken expansive rock mass, the support forces of the AFR including cable, bolt, hydraulic column and so on, the short cantilever beam of the AFR roof reaches a stable balanced state. At this time, the impact coefficient of mining on the working face is $\lambda_1 = 1$, and the support coefficient of broken expansive rock mass is $\lambda_2 = 1$. The critical value of support force along gob side of the AFR in the pre-splitting stabilisation zone can be obtained. The specific design basis for support force along gob side of the AFR in the pre-splitting stabilisation zone is in Eq. (2.36).

$$F = \frac{B}{1+B}\sigma_c^{(B-1)/B}A^{-1/B}\gamma^{1/B}(b-a)^{(1+B)/B} + \gamma ah + \frac{1}{2}h^2\tan\theta$$
$$+ q(b + h\tan\theta) - \sigma_t(b-a) - \left(b/d - 1 + 2\cos\theta\right)p_a$$
$$- qh\tan\theta - p_c(b-2l) \quad (2.36)$$

According to Eqs. (2.34)–(2.36), the support force design for AFR in different zones can be carried out including the parameter design of different types of bolts, cables, hydraulic columns, hydraulic supports, and other support materials inside the AFR.

2.3 Chapter Summary

(1) This chapter focuses on the traditional longwall mining method and conducts an analysis of the mining pressure behavior law during the entire mining period of the working face. Moreover, the movement law of the overlying rock and the mechanical state at various points during the adjacent mining face of the 121 mining method are thoroughly explored from three aspects: the force characteristics of the mine pillar, the movement law of the overlying rock structure, and the force state of the roadway support system.

(2) The rock damage produced in each stage of the mining process was examined from the standpoint of mining rock mechanics to elucidate the underlying causes for the occurrence of the aforementioned non-balance state in the 121 mining method. The mining damage invariant equation was proposed and solved, providing a foundation for balance control in mining activities. Based on balanced mining concept, combined with the innovative idea of transforming the

resistance to mine pressure into the utilization of mine pressure, and utilizing the rock mass broken expansion to fill underground space, a short cantilever beam structural model was established.

(3) The function equation of broken expansion and roadway broken expansion control equation of the collapsed rock mass in the AFR without mine pillars were obtained. After mining, the working face's balance state primarily takes three forms: the gob's automatic balance, the roadway's balance, the overall balance of several working faces. At the same time, the mechanical model of AFR roof zoning control is established. It provides a theoretical basis for the design of AFR control parameters.

References

He MC, Wang Q, Wu QY (2021) Innovation and future of mining rock mechanics. J Rock Mech Geotech Eng 13(1):1–21

Wang Q, Jiang B, Wang L, Liu BH, Li SC, Gao HK, Wang Y (2020) Control mechanism of roof fracture in no-pillar roadways automatically formed by roof cutting and pressure releasing. Arab J Geosci 13(6):274

Chapter 3
AFR Mining Method and Key Technologies

In order to achieve AFR without mine pillars, it is necessary to develop corresponding technologies according to balance mining theory. It mainly includes in-situ drilling and zoning technology for surrounding rock, directional roof cutting, high prestressed energy absorbing control, roadway gangue rib support technology. The specific content is shown in Fig. 3.1.

The various technologies cooperate with each other to form AFR mining method without mine pillars, as shown in Fig. 3.2. (a) This method utilizes in-situ drilling of surrounding rock to accurately partition the roadway roof, guiding roof cutting & support parameters design. (b) High prestressed energy absorbing support are used to assure the roadway roof's stability during the mining stages. (c) The gob roof is collapsed using mining pressure after determining the cutting direction, which severs the gob roof's physical link to the roadway roof quantitatively. The gob is filled with rock mass broken expansion, forming a stable gangue rib. (d) Maintain the AFR through the roadway gangue support system. Finally, this method eliminated the retention of mine pillars and automatically formed a roadway, while retaining the AFR for continued use in the next working face. The specific introduction of core technologies is as follows.

3.1 In-Situ Drilling and Zoning Technology

The core of in-situ drilling zoning technology for surrounding rocks is digital drilling testing. This method is a technology that utilizes digital drilling equipment to control, monitor, and analyze rock drilling parameters. Conduct strength testing and zoning of roadways surrounding rocks while drilling, as well as roof detecting. It is an important prerequisite for the design and optimization of pressure releasing and support schemes. As a leading link in the technology of the AFR, the in-situ drilling and zoning technology can provide the basic rock parameter basis for other key technologies.

Q. Wang et al., *Theory and Method of Automatically Formed Roadway in Underground Mining*, https://doi.org/10.1007/978-981-97-8082-2_3

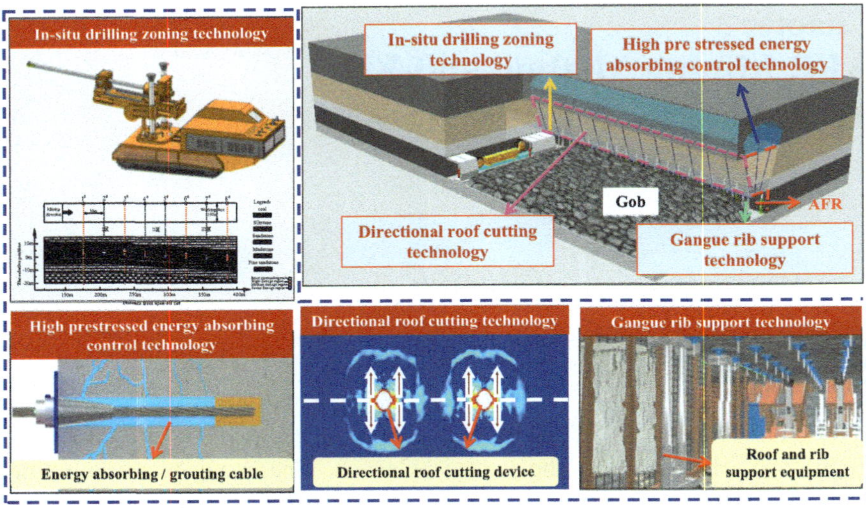

Fig. 3.1 Schematic diagram of key technologies for the AFR mining method

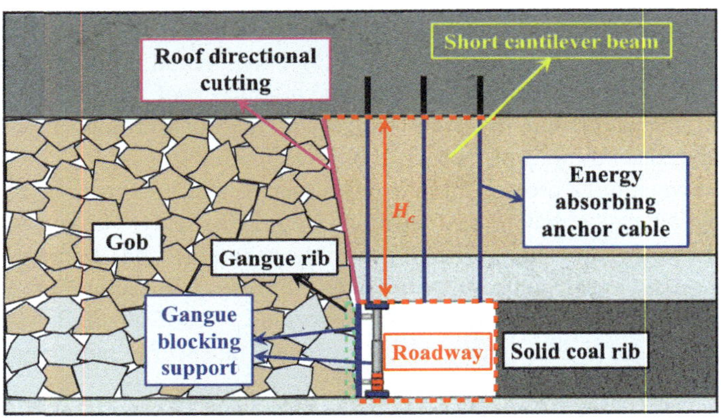

Fig. 3.2 Roadway cross-section diagram of AFR without mine pillars

3.1.1 *The Technology Principle*

The aim is to develop a quantitative link between drilling parameters and rock mass strength characteristics in order to accomplish in-situ collection of these values while drilling. The inversion model of rock mechanics parameters while drilling is obtained through rock cutting mechanics analysis. Based on this, to accurately analyze rock cutting mechanics, a rock mechanics parameter inversion model while drilling that integrates multiple types of digital drilling parameters is established. A new type of drill bit called the digital drilling analysis drill bit is designed, as shown in Fig. 3.3a.

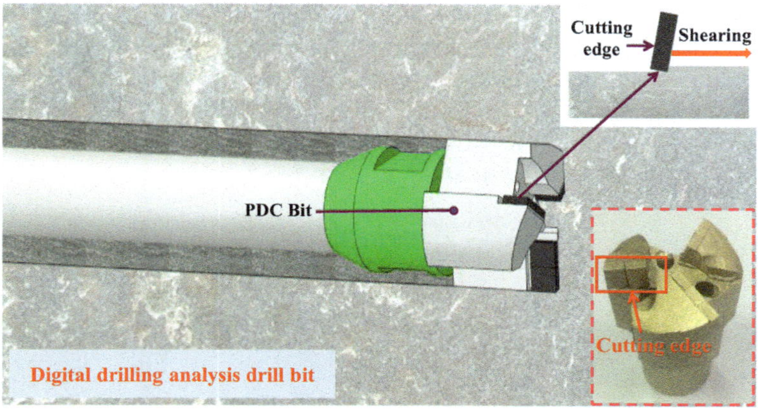

(a) Digital drilling analysis drill bit

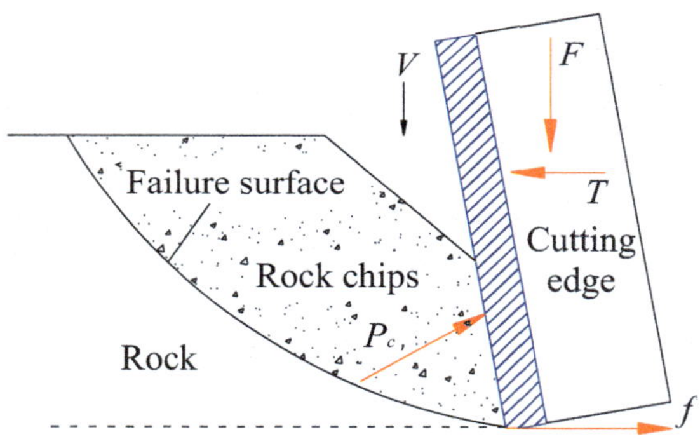

(b) Force analysis of drill cutting edge

Fig. 3.3 Digital drilling analysis drill bit and model analysis

The mechanical study of the digital drilling bit's cutting process was done in order to quantify the link between drilling parameters and the UCS of rock based on digital drilling analysis drill bits, as shown in Fig. 3.3b. The following forces are at work on the digital drill bit's cutting edge: (i) The force T applied by the digital drilling rig for breaking rock mass. The combined torque of the force T on all cutting edges of the drill bit to the center of the drill bit is the drilling torque M. (ii) The cutting edge of the drill bit is subjected to resistance P_c from the rock mass ahead. (iii) The drilling pressure F applied by the digital drilling rig. (iv) The frictional force f between the cutting edge and the rock mass at the bottom of the hole.

By analyzing the force state of the cutting edge of the drill bit, the work done by the digital drilling rig in rock mass includes torque work W_T and drilling pressure work W_F. The total energy consumed during the process of drilling rock mass is the sum of the energy E_c of the drill bit cutting the broken rock mass and the energy E_F

of the friction between the drill bit cutting edge and the rock mass at the bottom of the hole. Based on the principle of energy conservation and the relationship between drill bit drilling parameters, the inversion model of rock mass strength while drilling is established, as shown in Eqs. 3.1 and 3.2.

The energy consumed by the drill bit in cutting a unit volume of rock mass is defined as η_s. η_s is shown in Eq. 3.1.

$$\eta_s = \frac{2\pi NM - \pi\mu NF\left(2R_s - \frac{L_1^2+L_2^2+L_3^2}{L_1+L_2+L_3}\right) + FV}{\pi R_s^2 V} \tag{3.1}$$

A rock mass *UCS* parameter inversion model while drilling based on drill bits is established.

$$UCS = \alpha\eta_s + \beta = \alpha\frac{2\pi NM - \pi\mu NF\left(2R_s - \frac{L_1^2+L_2^2+L_3^2}{L_1+L_2+L_3}\right) + FV}{\pi R_s^2 V} + \beta \tag{3.2}$$

where α and β are the fitting coefficients under the assumption of a linear relationship between rock mass *UCS* and rock mass cutting energy η_s.

3.1.2 The Technical Performance Testing

(1) **Rock drilling test system**

In order to conduct research on rock mass digital drilling testing, a multi-function true triaxial rock drilling test system (TRD system) is developed, which includes drilling system, loading system, monitoring & control system and pressure chamber system, as shown in Fig. 3.4. It can achieve in-situ testing of roadway surrounding rock while drilling and obtain rock mass mechanics parameters.

(2) **Rock drilling test**

The TRD system is used to conduct indoor drilling tests on different rock masses. Drilling parameters V, N, F, and M of each rock strata of the digital drilling test are counted, and the rock cutting energy based on analytical drill bits η_s is calculated. Meanwhile, uniaxial compression tests are conducted on rocks to determine the uniaxial compressive strength R_c of each rock strata. The specific results are shown in Table 3.1.

(3) **Establishment of rock mass inversion model while drilling**

Based on the drilling test results of each rock strata in Table 3.1, draw a scatter plot for analyzing the relationship between rock cutting energy η_s and the *UCS* of rock, as shown in Fig. 3.5.

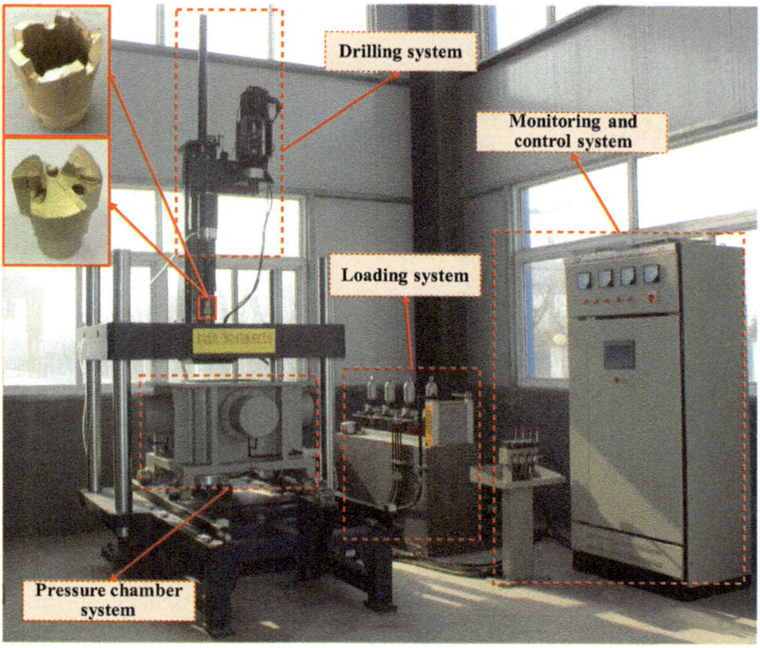

Fig. 3.4 Multi-function true triaxial rock drilling test system (TRD System)

Table 3.1 Statistics of digital drilling test and uniaxial compression test results

Specimen number	V (mm/min)	N (r/min)	M (N m)	F (kN)	η_s (MPa)	UCS (MPa)
S11	84.56	100.32	118.43	10.18	208.34	75.23
S12	86.06	100.07	4.03	0.13	9.11	–
S13	84.67	100.21	70.79	3.37	151.88	58.29
S21	86.71	100.20	5.78	0.91	5.81	10.65
S22	84.78	100.21	6.55	0.37	13.44	–
S23	84.37	100.27	48.49	2.52	102.30	35.45
S31	85.66	100.22	8.19	1.05	10.73	9.62
S32	86.65	100.26	5.32	0.28	10.89	–
S33	86.29	100.23	1.64	0.14	2.84	2.48
S41	86.35	100.32	30.75	1.93	60.12	19.98
S42	86.43	100.23	40.78	2.07	84.46	21.13
S43	85.53	100.33	21.73	1.23	44.24	18.99
S51	86.72	100.25	10.76	1.13	16.42	16.49
S52	86.46	100.24	0.48	0.05	0.74	1.59
S53	86.47	100.24	4.42	0.53	6.11	10.30

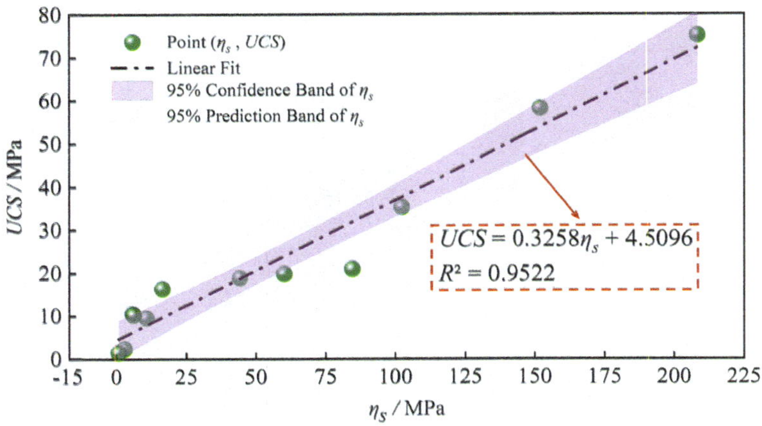

Fig. 3.5 Scatter plot of the relationship between rock cutting energy η_s and the *UCS* of rock

As shown in Fig. 3.5, the rock η_s obtained from digital drilling tests increases with the increase of the *UCS* of rock, and there is a high degree of linear response relationship between them. The linear regression statistical analysis method is used to obtain the quantitative relationship between rock η_s and rock *UCS*, as shown in Eq. (3.3). The linear goodness of fit R^2 is 0.9522, which shows the linear relationship between the η_s of rock and the *UCS* of rock is outstanding. The high degree of fitting proves the correctness of the linear relationship assumption in Eq. (3.2).

$$UCS = 0.3258\eta_s + 4.5096 \tag{3.3}$$

Therefore, the fitting coefficients α in Eq. (3.2) are 0.3258 and β are 4.5096, a rock *UCS* parameter inversion model while drilling based on an analytical drill bit as follows:

$$UCS = 0.3258\frac{2\pi NM - \pi\mu NF\left(2R_s - \frac{L_1^2+L_2^2+L_3^2}{L_1+L_2+L_3}\right) + FV}{\pi R_s^2 V} + 4.5096 \tag{3.4}$$

(4) **In-situ drilling testing and zoning method for surrounding rocks**

Based on the *UCS* inversion model for rock mass while drilling, a method for testing and zoning the mechanical parameters of the roadway surrounding rocks during drilling is established, as shown in Fig. 3.6. The specific steps are as follows:

1) Drilling parameters, including drilling speed V, torque M, drilling pressure F, and rotational speed N, are acquired in real-time throughout the whole drilling depth direction in the roadway using the on-site drilling system.
2) The equivalent mechanical parameter *UCS* of the surrounding rock in the entire drilling depth direction is inverted based on the rock mechanics parameter while

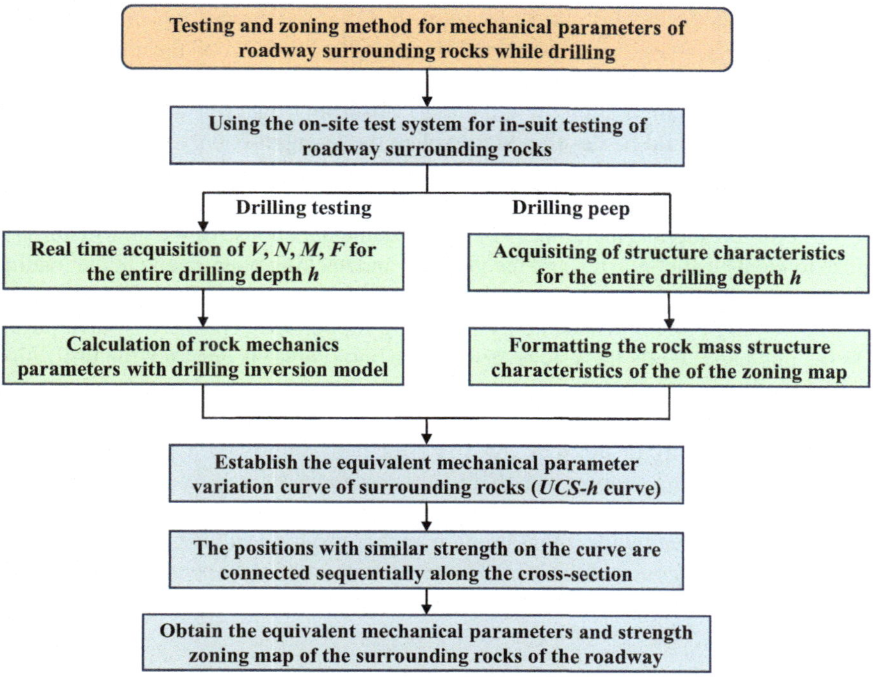

Fig. 3.6 Testing and zoning method for mechanical parameters of roadway surrounding rock while drilling

drilling inversion model and the drilling parameters obtained from testing. The *UCS* variation curve with drilling depth h (*UCS-h* curve) is drawn.

3) To get *UCS-h* curves for various drilling places and orientations in the same portion of surrounding rocks, repeat Step (1) and Step (2). An equivalent strength zoning map of the surrounding rocks is created by connecting the points on the curve with equal strengths along the cross-section. The AFR mining technology may be designed with a parameter basis and guidance due to the achievement of real-time in-situ testing and zoning of roadway surrounding rocks.

3.2 High Prestressed Energy Absorption Control Technology

In regard to the bolt or cable material commonly used in the field (hereinafter referred to as common material), the elongation decreases with the increasing strength. This leads to the common materials cannot have the characteristics of high strength and high elongation at the same time. Since the energy is related to both force and displacement, the ability of traditional materials to absorb the energy of surrounding rock is limited. Meanwhile, in order to ensure that the supporting materials have sufficient safety margin, the prestress is generally 30–50% of the material yield strength.

Therefore, new high-strength and energy absorbing material with the following characteristics should be developed to realize the strength-energy coupled support: (1) The yield strength of the new material is more than double that of the common support materials. (2) The yield strain of the new material is more than 4 times that of the common support material. (3) The mechanical response of the new material is close to ideal elastic–plastic, and the yielding and necking phenomena in the loading process are eliminated as much as possible.

The new high-strength and energy absorbing support material has two advantages. On the one hand, while the safe reserve of the bolt (cable) is ensured, the utilization rate of the material strength is improved. The prestress can be 70–90% of its yield strength to compensate the stress unloaded by the roof rock due to excavation. The self-bearing capacity of the rock mass is effectively improved. On the other hand, the energy absorbing support can effectively absorb the energy released during rock mass deformation. It can effectively control the disasters such as rock bursts in deep roadways. Meanwhile, high-strength anchor grouting support is required in the broken surrounding rock of the roof. The systems work together with the roof cutting technology to ensure the stability of the surrounding rock in the mining process.

3.2.1 *The New Support Material*

Based on the above objectives, high prestressed energy absorbing bolt and cable are created from this innovative support material, which has a significant elongation ratio, so as to absorb released energy in the AFR. The purpose of the laboratory testing is to determine the mechanical performance of the new bolt (cable) in comparison to standard bolts, as well as the degree of coupling. Figures 3.7 and 3.8 depict the new energy absorbing bolt and cable (hereinafter referred to as EAB and EAC).

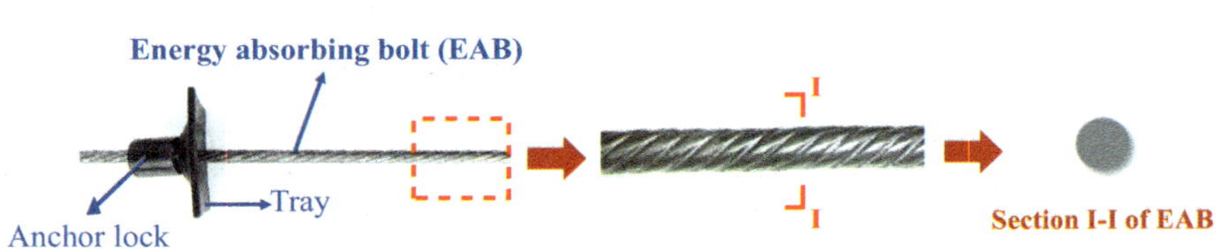

Fig. 3.7 The new energy absorbing bolt. Reprinted from Ref. Wang et al. (2022), Copyright © 2022 Elsevier Ltd.

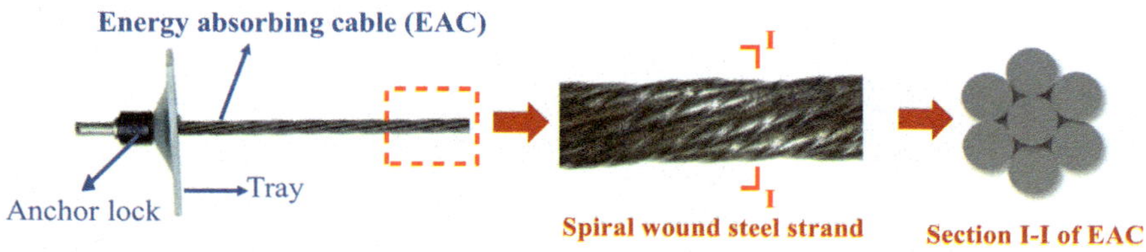

Fig. 3.8 The new energy absorbing support cable. Reprinted from Ref. Wang et al. (2022), Copyright © 2022 Elsevier Ltd.

3.2.2 The Material Performance in Static Mechanical Tests

Static tension tests are conducted to elucidate the mechanical characteristics of energy absorbing bolt (cable). Table 3.2 presents the test scheme. In this table, CB refers to common bolt. The test system utilized in this work is displayed in Fig. 3.9.

Figure 3.10 displays the load-elongation ratio curves and failure scenarios for EAB-S, EAC-S, and CB-S. Figure 3.11 displays the mechanical and energy absorbing property index statistics.

One may assess the specimens' mechanical properties, deformation, and energy absorbing qualities using Figs. 3.10 and 3.11.

(1) **Mechanical characteristics analysis**

   (i) Support strength: 230.8 kN and 397.4 kN, respectively, are the breaking forces of EAB-S and EAC-S. The energy absorbing bolt and cable have a high-strength supporting capability, as evidenced by the maximum loads of EAB-S and EAC-S, which are 1.63 and 2.80 times that of CB-S, respectively, and 141.8 kN for CB-S.

Table 3.2 The static tension test scheme of energy absorbing material

Test type	Test number	Test specimen	Length/m	Loading mode	Test system parameter
Static tensile	CB-S	Common bolt	1.5	Displacement-controlled uniform tension	(1) Maximum tensile load: 500kN (2) Tensile displacement loading rate: 10 mm/min
	EAB-S	Constant resistance energy absorbing bolt	1.5		
	EAC-S	Constant resistance energy absorbing cable	1.5		

Reprinted from Ref. Wang et al. (2022), Copyright © 2022 Elsevier Ltd.

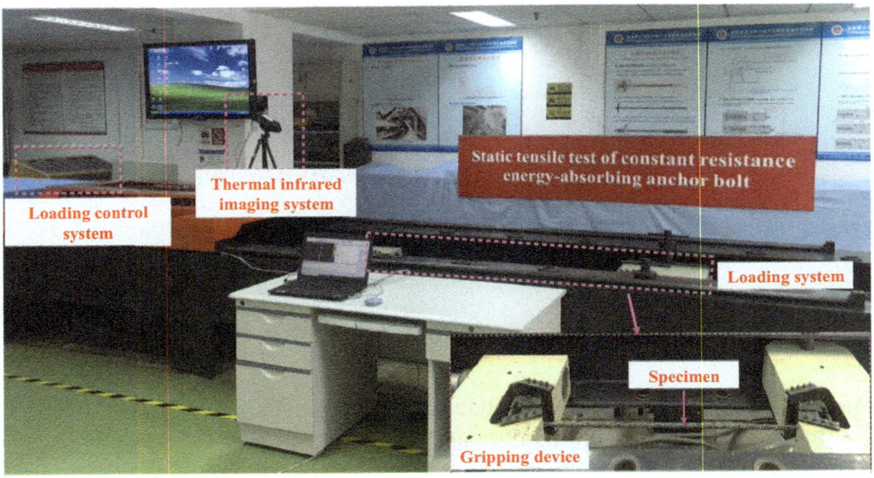

Fig. 3.9 New material static tension performance testing system. Reprinted from Ref. Wang et al. (2022), Copyright © 2022 Elsevier Ltd.

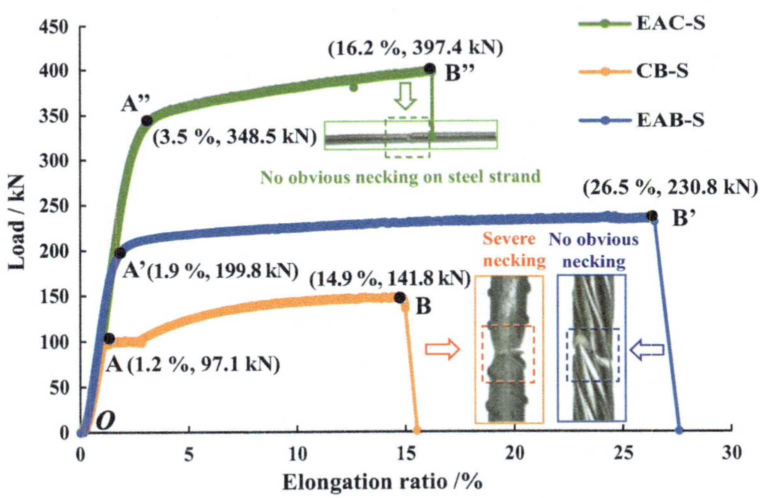

Fig. 3.10 Load-elongation ratio curve of bolts (cables). Reprinted from Ref. Wang et al. (2022), Copyright © 2022 Elsevier Ltd.

(ii) Constant resistance supporting capacity: By dividing the difference between the maximum and minimum loads by the maximum load during the constant resistance stage, one may get the constant resistance fluctuation rate, which symbol is represented by μ. There is less fluctuation in

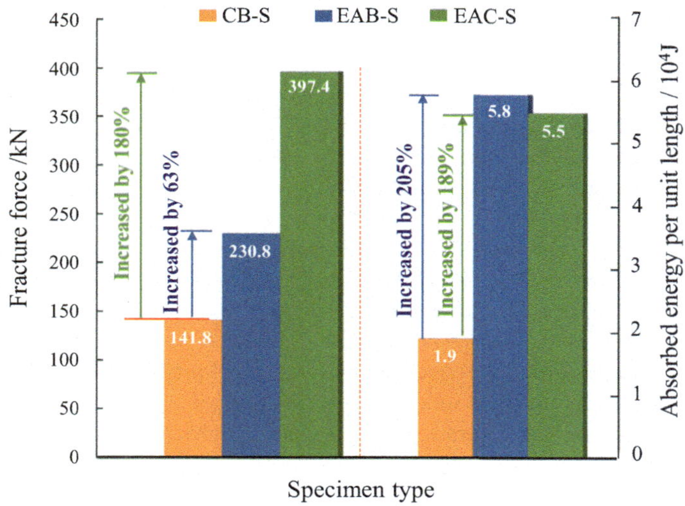

Fig. 3.11 Comparison of static performance of bolts (cables). Reprinted from Ref. Wang et al. (2022), Copyright © 2022 Elsevier Ltd.

reinforcing resistance with a lower μ value. The constant resistance fluctuation rates of EAB-S and EAC-S are 15.5% and 14.0%, respectively, which are 66.3% and 69.5% lower than those of CB-S. As a result, the energy absorbing bolt and cable can maintain constant resistance.

(2) **Analysis of deformation and energy absorbing properties**

(i) Properties of deformation: There is no noticeable necking phenomena, and the elongation ratios of EAB-S and EAC-S are 26.5% and 16.2%, respectively. CB-S has an elongation ratio of 14.9%. The maximum elongations of EAC-S and EAB-S are 1.09 and 1.77 times, respectively, more than those of the latter. It indicates that the energy absorbing bolt (cable) possesses significant elongation properties.

(ii) Capabilities to absorb energy: The energy absorbed by the bolt (cable) per unit length is the region in Fig. 3.10 that is bounded by the load-elongation ratio curve and the rectangular coordinate axis. For EAB-S, EAC-S, and CB-S, the absorbed energy per unit length is 5.8×10^4 J/m, 5.5×10^4 J/m, and 1.9×10^4 J/m, in that order. It is evident that the energy absorbing bolt (cable) has an advantage over the CB-S in terms of absorbing energy generated by the rock deformation from the absorbed energy per unit length of EAB-S and EAC-S, which are 3.05 and 2.89 times that of CB-S, respectively.

3.2.3 The Material Performance in Dynamic Mechanical Tests

Dynamic impact tests are conducted to clarify the mechanical characteristics of energy absorbing bolt and cable. Table 3.3 presents the test scheme. The test system utilized in this work is displayed in Fig. 3.12.

The breaks occurred at the 22nd and 64th times, respectively, for EAB-D and EAC-D. After the tenth time, CB-D broke. Furthermore, it can be noted that there is a necking phenomenon in CB-D, but not at all in the broken parts of EAB-D and EAC-D. The dynamic impact test employs the equal impact weight and height loading approach, and the impact energy per drop is 1.0×10^4 J. The single impact force and displacement for the tested specimens are displayed in Fig. 3.13. Figure 3.14 displays the energy absorbing property index and dynamic impact resistance parameters.

The impact resistance parameters and energy absorbing characteristics are displayed in Figs. 3.13 and 3.14.

(1) **Impact resistance property analysis**

The specimens' single impact displacement progressively diminishes with the accumulation of impact times, as shown in Figs. 3.13 and 3.14. The average single impact force of EAB-D and EAC-D increased by 12.4% and 53.6%, respectively, in comparison to CB-D. For EAB-D and EAC-D, the average single impact displacement is reduced by 39.5% and 86.7%, respectively. With the same dynamic impact energy, the energy absorbing bolt (cable) deforms less, suggesting that it has an advantage in terms of impact resistance.

(2) **Energy absorbing and deformation property analysis**

(i) Deformation property analysis: In the initial hit, there was a significant displacement due to the slip between the load unit and the specimens. As a result, the secondary coordinate axis' accumulation of single impact displacement in Fig. 3.13 is used to compute the specimens' elongation. EAB-D and EAC-D had elongation ratios that are 1.45 and 1.04

Table 3.3 The dynamic test scheme of constant resistance energy absorbing material

Test type	Test number	Test specimen	Length/m	Loading mode	Test system parameter
Dynamic impact	CB-D	Common bolt	3.0	Impact by drop hammer with equal weight and equal height	(1) Drop hammer weight: 1000 kg (2) Drop height: 1.0 m
	EAB-D	Constant resistance energy absorbing bolt	3.0		
	EAC-D	Constant resistance energy absorbing cable	3.0		

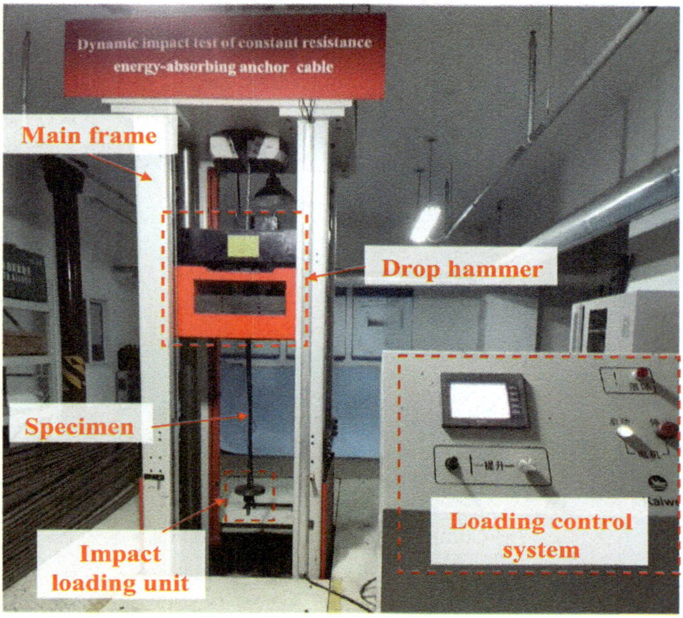

Fig. 3.12 New material dynamic performance testing system. Reprinted from Ref. Wang et al. (2022), Copyright © 2022 Elsevier Ltd.

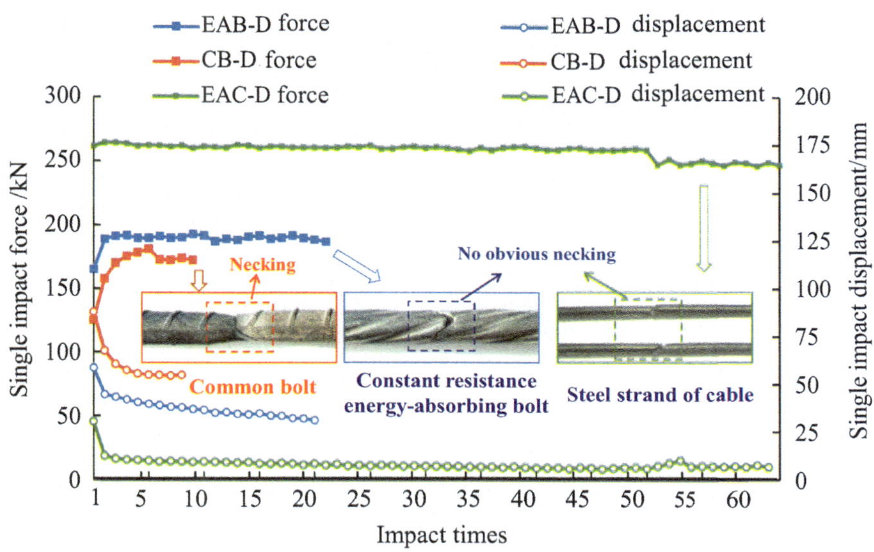

Fig. 3.13 The impact load/displacement curve. Reprinted from Ref. Wang et al. (2022), Copyright © 2022 Elsevier Ltd.

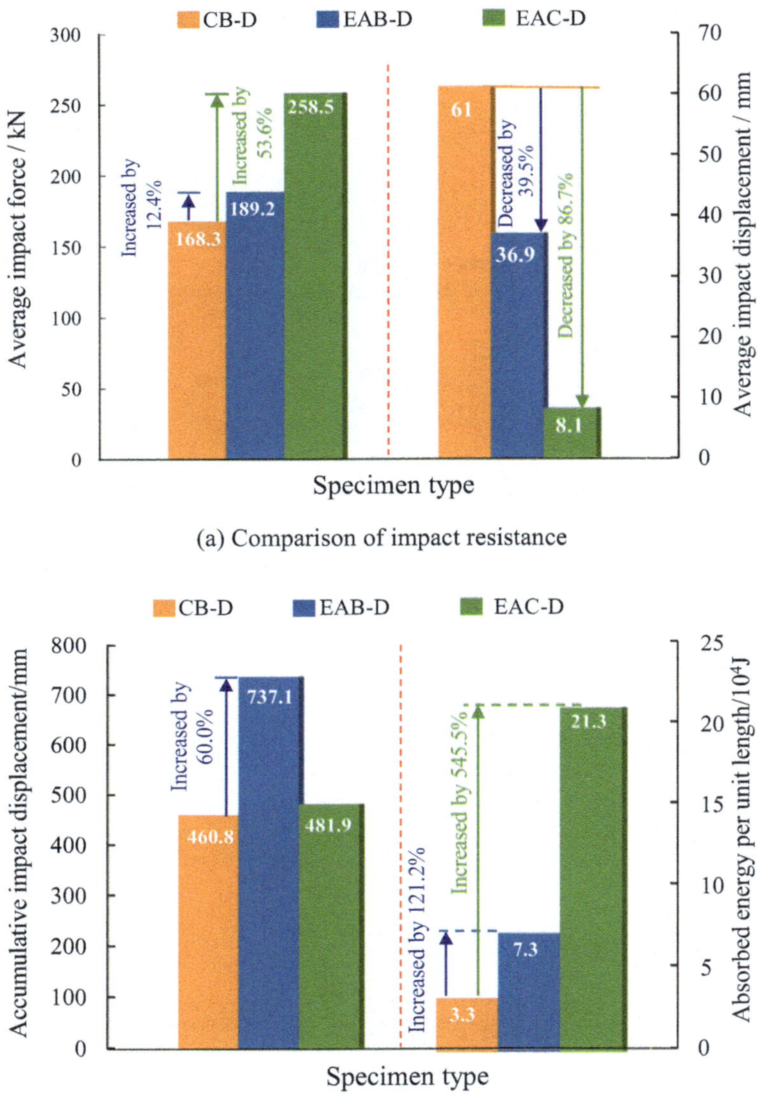

(a) Comparison of impact resistance

(b) Comparison of displacement and absorbing energy

Fig. 3.14 Comparison of the specimens' dynamic performance. Reprinted from Ref. Wang et al. (2022), Copyright © 2022 Elsevier Ltd.

times greater than CB-D, respectively. The findings indicate that energy absorbing bolts have the best overall deformation performance under dynamic impact, followed by energy absorbing cables and common bolts.

(ii) Examination of the property that absorbs energy: EAB-D, EAC-D, and CB-D absorb 7.3×10^4 J/m, 21.3×10^4 J/m, and 3.3×10^4 J/m of energy per unit length, respectively. For EAB-D and EAC-D, the energy absorbed per unit length is 2.18 times and 6.4 times that of CB-D, respectively. The energy absorbing bolt (cable), according to the results, has benefits in terms of energy absorption.

3.3 Directional Roof Cutting Technology

The theory of balancing mining states that by converting the roof long cantilever beam construction to a short cantilever beam, the stress transfer between the gob and the roadway roof is cut off. By using mining stress & rock mass broken expansion properties in the gob, it is possible to achieve directed collapse of the gob roof and the automatic formation of roadway. Therefore, new method is required to achieve roof cutting so as to precisely control gob roof collapse height. Bidirectional energy gathering blasting technique has been presented as a directional roof cutting technology by using the strong compressive σ_c and weak tensile σ_t characteristics of the rock mass, as shown in Fig. 3.15.

3.3.1 The Technology Principle

A directional roof cutting of surrounding rock is achieved through the energy gathering effect of a bidirectional energy gathering blasting device in a set direction. The blast hole is filled with explosives using a bidirectional energy gathering blasting device, that is, the explosives roll is first loaded into the device. The device is placed in blast hole according to design direction. After the explosive is detonated, the device produces an instantaneous suppression effect on the detonation products. The two

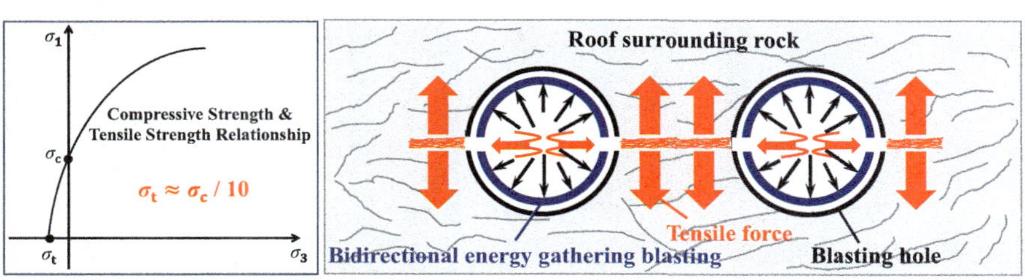

Fig. 3.15 The schematic diagram of bidirectional energy gathering blasting

linearly distributed rows of energy-gathering holes provide priority for instantaneous pressure relief space, forming high-energy flow at the energy-gathering holes, which concentrate on the corresponding hole walls and generate radial initial cracks. The high-speed detonation gases continue to be preferentially released from the energy-gathering holes following the previous detonation products. This allows the gas to create a strong "gas wedge" inside the radial initial crack, which keeps the crack from stabilizing. This provides a new free surface for stress waves, generating reflected tensile stress concentration in the vertical set direction, accelerating the directional propagation of cracks, resulting in tensile cracking of the rock mass. The specific content is shown in Fig. 3.15.

3.3.2 *The Directional Cutting Performance Test*

The bidirectional energy gathering blasting device is different from the previous "point" or "line" shaped energy gathering devices. The details are as follows (Fig. 3.16).

(1) The design strength of this device varies depending on the rock mass type, especially when processed from pipelines with set strength. The bidirectional energy gathering blasting device plays a role in the function of energy gathering and tensioning. On the one hand, a specific strength must be attained so as to lessen the harm that explosion products cause to the reserved surrounding rock. It creates the ideal circumstances for both localized focused stress and

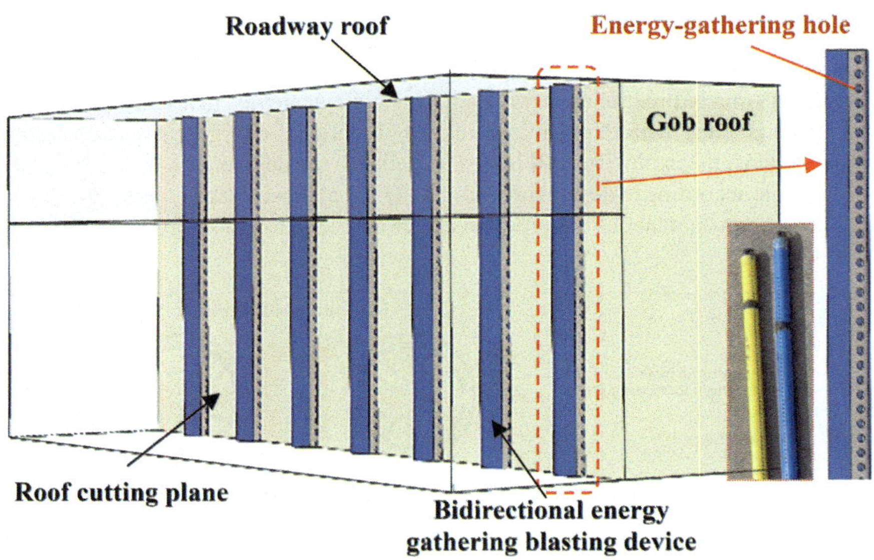

Fig. 3.16 Bidirectional energy gathering blasting device

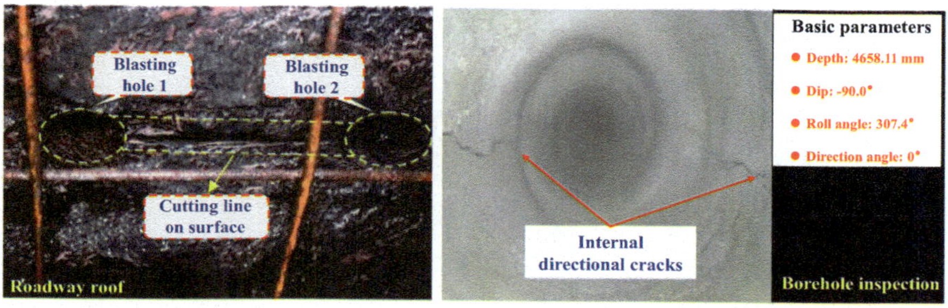

Fig. 3.17 The effect diagram of tension blasting

overall uniform compression of the surrounding rock. To minimize the energy consumption operating on the bidirectional energy gathering blasting device, the device's strength must not be excessively high.

(2) The diameter of the device varies with the diameter of the blast hole, and is determined based on its coupling relationship with the blast hole.
(3) Energy can be collected in any two directions, and the angle of energy accumulation depends on the shape of the blasting engineering contour line, such as 90°, 135°, 180°, etc.
(4) The design length of the bidirectional energy gathering blasting device is generally 0.5–1.0 m, which can be combined according to needs during application to facilitate on-site operation. The schematic diagram of the bidirectional energy gathering blasting device is as follows.

Due to the buffering effect of the pipe wall and the preferential unloading of detonation products from the energy-gathering holes, the stress effect sharply decreases, reducing the damage of detonation products to the hole wall. Therefore, it suppresses the development of non-directional cracks. When several blasting holes are simultaneously detonated, a superimposed stress field will be generated between the holes, and the tensile stress between the holes will increase. If the spacing between blasting holes is appropriate, adjacent cracks can be connected to form a smooth directional cutting line on surface, achieving precise control of blasting roof cutting, as shown in Fig. 3.17.

The bidirectional energy gathering blasting device can achieve the goal of quantitative directional roof cutting, providing a guarantee for the implementation of the AFR mining method.

3.4 Roadway Gangue Rib Support Technology

At present, the common gangue blocking support technology for retaining roadways is U-shaped steel, hydraulic column and steel mesh, as shown in Fig. 3.18. Prior to the gob roof collapsing, there will be a significant vertical pressure buildup on the

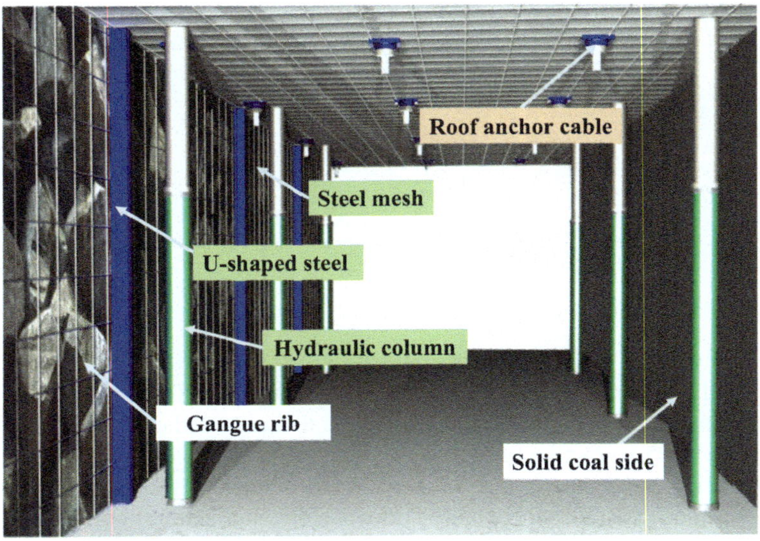

Fig. 3.18 The U-shaped steel for supporting roadway rib

AFR, and high dynamic pressure will develop during the gob roof collapse. Under action of the above factors, the traditional gangue blocking support—U-shaped steel support is prone to deform instability.

Therefore, it is necessary to study a support technology that can resist large pressure, while forming a gangue rib and ensuring stability. In order to address the above issues, two types of gangue rib support technologies are proposed, which mainly includes gangue rib support by confined concrete technique and gangue rib support by hydraulic support technique. It refers to the use of the above technologies to achieve safety control of the gob side of the AFR, ensuring roadway stability during the formation stage.

3.4.1 Gangue Rib Support by Confined Concrete Technique

Confined concrete is a composite support structure in which the concrete is poured into the confined materials such as steel tube. Confined concrete has a bearing capacity under axial compression that is 2.6 times greater than that of U-shaped steel with the same section steel composition. This construction may efficiently regulate the surrounding rock's stability around AFR, guaranteeing safe and effective regulation of the surrounding rock's stability. In order to determine structure bearing capacity under eccentric compression, structures mechanical tests under different eccentric distance are carried out to reasonably design the support structure.

(1) **Test scheme design**

To analyze bearing capacity of the roof and rib support structure under the condition of eccentric pressure, U29 steel commonly used on site is selected as the contrast object. The eccentric contrast test of confined concrete support structure is carried out to conduct reasonable selection. The confined concrete section is designed according to the principle that the steel content of the U-shaped steel section is equal. Square steel tube with side length of 160 mm and wall thickness of 6 mm is selected for the square confined concrete, filled with LC40 lightweight aggregate concrete (hereinafter referred to as SQLC) and C40 ordinary concrete (hereinafter referred to as SQC). Circular steel tube with a diameter of 159 mm and a wall thickness of 8 mm is selected for circular confined concrete (CLC), filled with LC40 lightweight aggregate concrete.

The purpose of the test schemes with varying eccentric distances is to investigate how the eccentric distance affects the support structures' bearing behavior. The eccentric distance is designed as 40, 80 and 120 mm, as shown in Table 3.4. The test specimens are divided into US specimens, CLC specimens, SQC specimens and SQLC specimens, as shown in Fig. 3.19. The test specimens are numbered as X–Y. X represents the cross-sectional form including US, CLC, SQC and SQLC, and Y is the eccentric distance of the test specimens including 40, 80 and 120 mm. Taking U29-40 as an example, it indicates that the cross-section is U29 steel and the eccentric distance is 40 mm.

(2) **Loading and monitoring scheme**

The two ends of the test specimen are connected by the bearing plate, and the eccentric load is applied to the test specimen through the loading cylinder, as shown in Fig. 3.20. As load reaches 80% of predicted ultimate load, the loading rate is set to 0.2 kN/s.

Table 3.4 Size of test piece

Test group	Cross-section	Test number	Eccentric distance/mm
I	U-shaped steel	US-40	40
		US-80	80
		US-120	120
II	Circular confined concrete	CLC-40	40
		CLC-80	80
		CLC-120	120
III	Square confined concrete	SQC-40	40
		SQC-80	80
		SQC-120	120
		SQLC-40	40
		SQLC-80	80
		SQLC-120	120

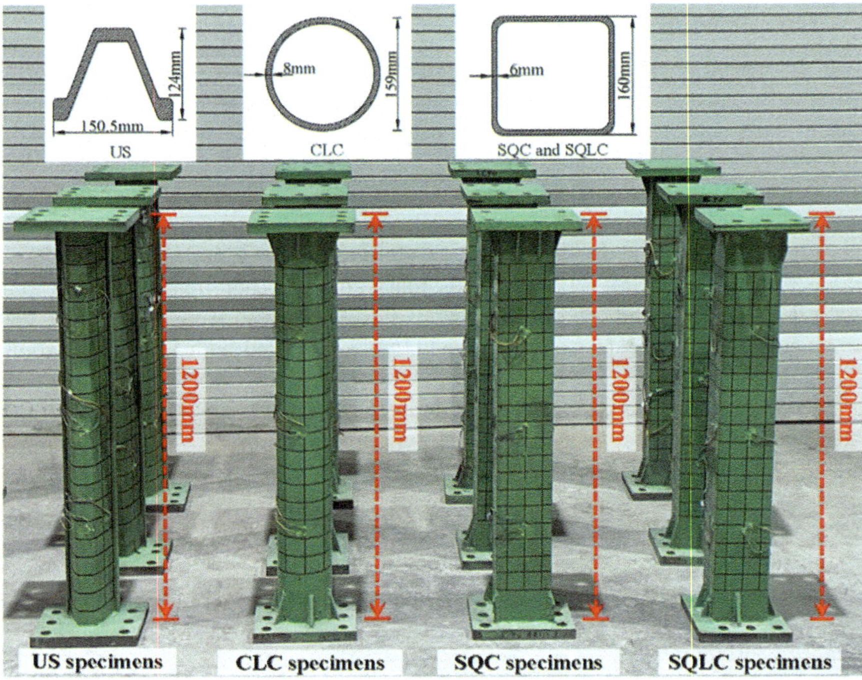

Fig. 3.19 Test specimens and its parameters

Displacement sensors are arranged at 1/4, 1/2 and 3/4 heights of the test specimen to monitor the deformation. The axial load and displacement on the top of the test specimen are recorded.

(3) **Test result analysis**

1) Analysis of bearing capacity and axial deformation characteristics

The axial stiffness is defined as the curve slope in the elastic deformation stage to analyze the axial deformation characteristics. Difference rate of ultimate bearing capacity between confined lightweight and ordinary concrete specimens is defined as $\delta_N = (N_{CL} - N_C)/N_C$. Difference rate of axial deformation stiffness is defined as $\delta_{AR} = (A_{CL} - A_C)/A_C$. N_i and A_i respectively represent the ultimate bearing capacity and axial stiffness of the test specimen, $i = CL$ and C respectively represent SQLC specimen and SQC specimen. The ultimate bearing capacity and axial stiffness of the specimens are shown in Figs. 3.21 and 3.22.

The analysis on Figs. 3.21 and 3.22 shows:

(a) When the eccentric distance is 40 mm, 80 mm and 120 mm, the ultimate bearing capacity of confined lightweight concrete specimens is increased by 124.8%, 97.2% and 95.3% respectively compared with the US specimens, and the axial

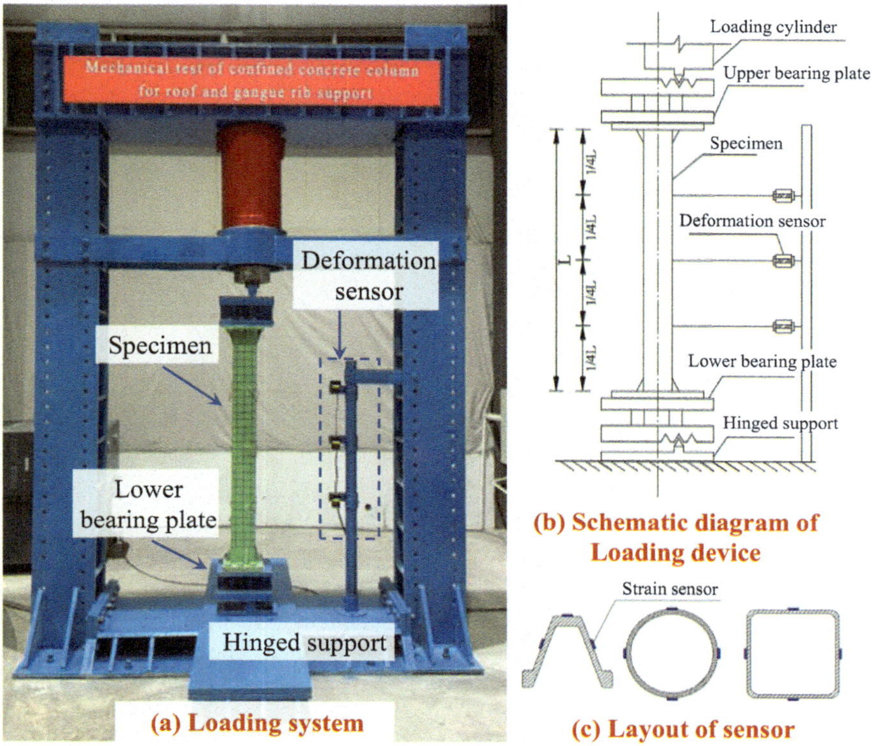

Fig. 3.20 Test loading device and design diagram

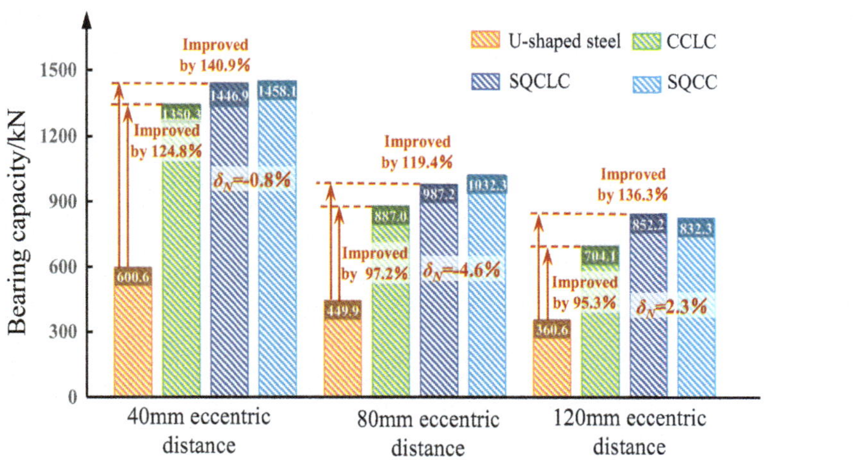

Fig. 3.21 Comparison of ultimate bearing capacity of different specimens

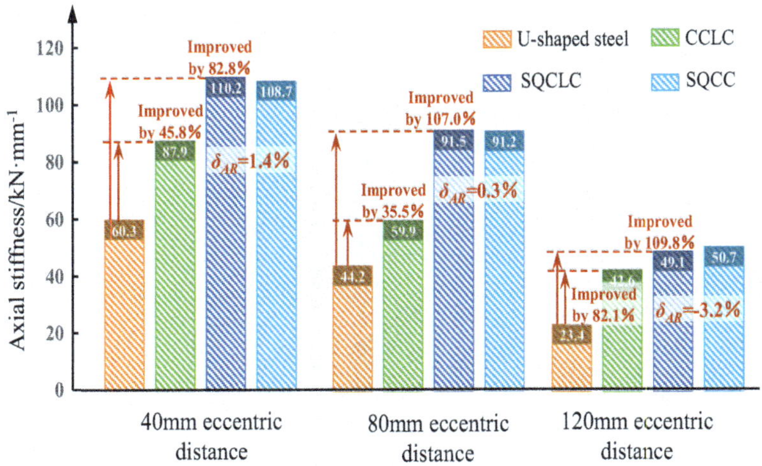

Fig. 3.22 Comparison of axial stiffness of different specimens

stiffness is increased by 45.8%, 35.5% and 82.1% respectively. The results show that confined lightweight concrete specimens have higher bearing capacity and axial stiffness under different eccentric distance. The support structure of confined lightweight concrete can provide vertical support to the roof more effectively.

(b) When the eccentric distance is 40 mm, 80 mm and 120 mm, the difference rates of ultimate bearing capacity are − 0.8%, − 4.6% and 2.3% respectively, and the difference rates of axial stiffness are 1.4%, 0.3% and 3.2% respectively. The results show that the ultimate bearing capacity and axial stiffness of SQLC and SQC specimens are basically the same under different eccentric distance.

(c) The ultimate bearing capacity of CLC-40 and SQLC-40 specimens is reduced by 47.9% and 41.1% respectively compared with CLC-120 and SQLC-120 specimens, and the axial stiffness is reduced by 51.5% and 55.4% respectively. The results show that the ultimate bearing capacity and axial stiffness of CLC and SQLC specimens obviously decrease with the increasing eccentric distance. In the design of AFR, the lateral load of roof and rib support structure should be reduced as much as possible to weaken the influence of eccentric distance on the bearing capacity of support structure.

2) Analysis of load-lateral deflection curves

The load-lateral deflection curves are shown in Fig. 3.23 to analyze the lateral resistance of the specimens. The ratio of the vertical load in the elastic stage to the mid span deflection is defined as the lateral stiffness. The comparison on the lateral stiffness of different test specimens is shown in Fig. 3.24. The difference rate of lateral stiffness between confined lightweight and ordinary concrete specimens is defined δ_{LR}, which is the ratio of the difference between the lateral stiffness of SQLC specimen and SQC specimen with the lateral stiffness of SQC specimen.

The analysis on Figs. 3.23 and 3.24 shows:

(a) The specimens similarly undergo elastic deformation stage, elastic–plastic deformation stage and instability failure stage. The load increases linearly with the lateral deflection in the elastic deformation stage. The load increases slowly until the specimen reaches the ultimate bearing capacity in the elastic–plastic deformation stage. The load decreases rapidly with the increasing lateral deflection until the specimen fails due to the influence of the second-order effect in the instability failure stage.

(b) Compared with the US specimens loaded with the same eccentric distance, the lateral stiffness of CLC-40, CLC-80 and CLC-120 specimens increased by

Fig. 3.23 Load lateral-deflection curves

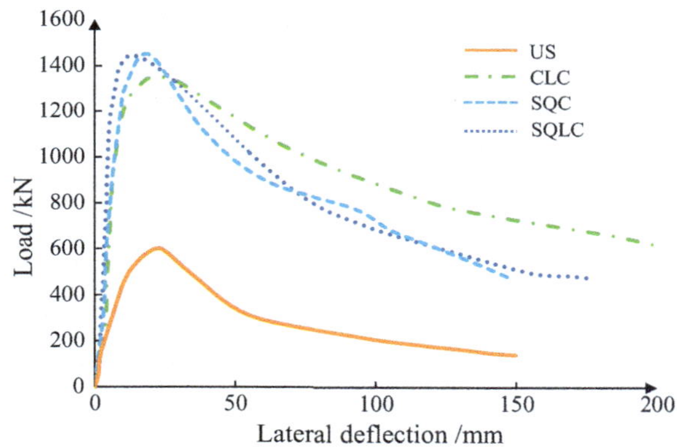

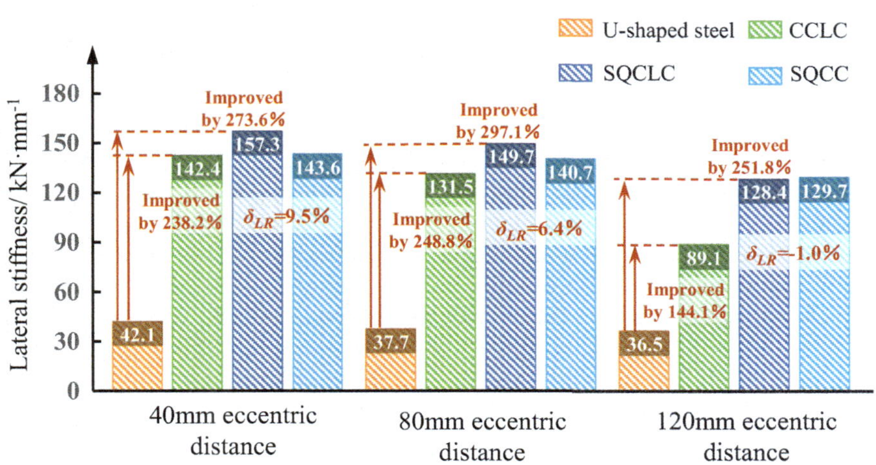

Fig. 3.24 Comparison on the lateral stiffness of different test pieces

238.2%, 248.8% and 144.1%, and the lateral stiffness of SQLC-40, SQLC-80 and SQLC-120 specimens increased by 273.6%, 297.1% and 251.8%. The results show that the lateral stiffness of CLC and SQLC specimens is significantly better than that of US specimens under the bias load. Therefore, the confined light concrete support structure has better stability in the process of roadway support.

(c) Compared with SQC specimens loaded with the same eccentric distance, the difference rates of lateral stiffness of SQLC-40, SQLC-80 and SQLC-120 specimens are 9.5%, 6.4% and − 1.0% respectively. The results show that the difference rate between the lateral bending stiffness of SQLC and SQC specimens is within 10%. With the increase of eccentric distance, the lateral stiffness of SQLC and SQC specimens is gradually close to each other. When the eccentric distance reaches 120 mm, the lateral stiffness of both specimens is basically the same.

3.4.2 Gangue Rib Support by Hydraulic Support Technique

The roof and rib support equipment is the core equipment of the gangue rib support by hydraulic support technique, as shown in Fig. 3.25. The roof and rib support equipment is composed of the top beam, bottom beam, push plate, and column. These components may offer a significant amount of supporting force to prevent roof deformation and preserve yield pressure to adjust to the specific deformation of the basic roof. The roof and rib support equipment can provide a vertical support force of 4000 kN by top beam and a lateral support force of 218 kN by push plate. Combination of two parts of the structure can combine multiple retaining structures, transforming the pressure of a single support component into collective bearing, which effectively alleviates the impact of the support.

The support technology adopts the following construction techniques on site. After mining and before the roof collapses, roof and rib support equipment support system is set up to installed on the side of the gob. After the cutting roof collapses, these composite support structures block the gangue. As the working face progresses, the gangue blocking board and temporary support equipment moves forward, the gob-side gangue gradually becomes compacted, and a gangue rib eventually forms. Additionally, the temporary roof-cutting support can also withstand the significant amount of vertical pressure from the gob area roof on the roadway before collapse and the significant amount of dynamic pressure during the collapse. These devices provide effective support for the gob side rib and roadway roof.

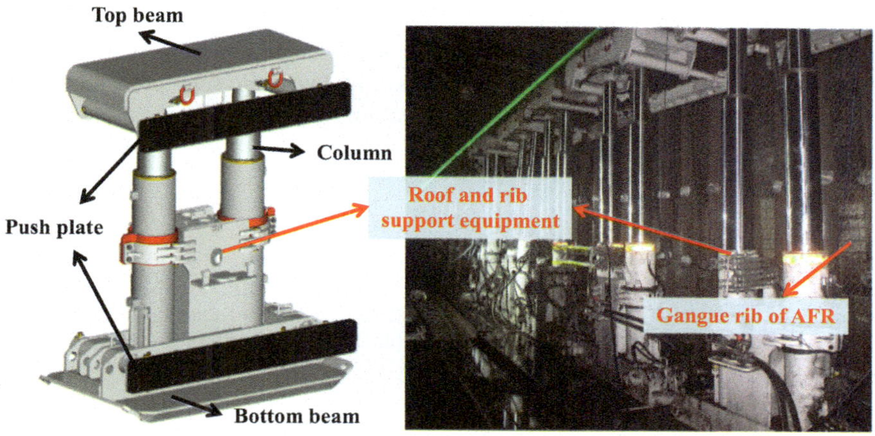

Fig. 3.25 Roof and rib support equipment support system

3.5 Chapter Summary

(1) The corresponding technologies are developed according to balance mining theory in this chapter. It mainly includes in-situ drilling and zoning technology for surrounding rock, directional roof cutting, high prestressed energy absorbing control, roadway gangue rib support technology.

(2) The core of in-situ drilling and zoning technology for surrounding rocks lies in digital drilling testing. This technology utilizes digital drilling equipment to control, monitor, and analyze rock drilling parameters. It can be used for testing and zoning the strength of roadway surrounding rocks while drilling, as well as detecting roof rock strata. It is an important prerequisite for the design and optimization pressure releasing & support schemes.

(3) The directional roof cutting technology utilizes the strong compressive and weak tensile properties of rock mass to develop a bidirectional energy gathering blasting device. Directional roof cutting is carried out to set the collapsed direction and range of rock mass, which is one of the core technologies for the AFR mining method.

(4) On the roof of the AFR, the high prestressed energy absorbing control technology is a support technique used to guarantee the stability of the roadway roof during mining process. Large elongation ratio bolts and cables with excellent strength and energy absorbing are created using this innovative support material. This type of support has good energy absorption control characteristics under static and dynamic conditions.

(5) The roadway gangue rib support technology is a support technology carried out on the gob side of the AFR. It mainly targets the high vertical pressure generated above the roof, and considering the impact of strong dynamic pressure during the collapse process on the gangue rib side, as well as the lateral compression

and impact effects of the collapsed gangue. Therefore, two types of gangue rib support technologies, which mainly includes gangue rib support by confined concrete technique and gangue rib support by hydraulic support technique are proposed. It can ensure the safety and stability of the gob side in the AFR.

Reference

Wang Q, Xu S, Xin ZX, He MC, Wei HY, Jiang B (2022) Mechanical properties and field application of constant resistance energy-absorbing anchor cable. Tunn Undergr Space Technol 125:104526

Chapter 4
Numerical Simulation Test of AFR Mining Method

The mining pressure is the primary source of the significant deformation of the roadway surrounding rocks, mine safety incidents, and other problems. The AFR mining method is a new type of mining pressure relief and roadway forming technology, which has certain advantages in reducing surrounding rock stress and maintaining roadway stability. This chapter uses numerical simulation means to conduct a comparative study between different types of AFR mining methods and traditional methods, in order to clarify the advantages of AFR mining methods and the impact of different factors on AFR surrounding rock control.

4.1 The AFR with Shallow Buried Depth

A typical shallow buried working face is used as the engineering backdrop in this study to clarify the stress fluctuation of the roadway and the control mechanism of the surrounding rock in AFR mining method and traditional methods. Based on real on-site working circumstances, a numerical calculation model is constructed. It is suggested to use the quantitative evaluating indices of the surrounding rock stress. Four kinds of numerical testing are methodically carried out. The specific content is as follows.

4.1.1 Research Object

Ningtiaota Mine is located in China's Shaanxi Province's Shenmu County. This mine has a planned production capacity of 12 million tons per year. Situated in the 2# coal seam, Ningtiaota Mine's S1201-II working face spans 280 m in dip length and 2344 m in strike length, as shown in Fig. 4.1. The average thickness of coal seam is 4.1 m, and

Q. Wang et al., *Theory and Method of Automatically Formed Roadway in Underground Mining*, https://doi.org/10.1007/978-981-97-8082-2_4

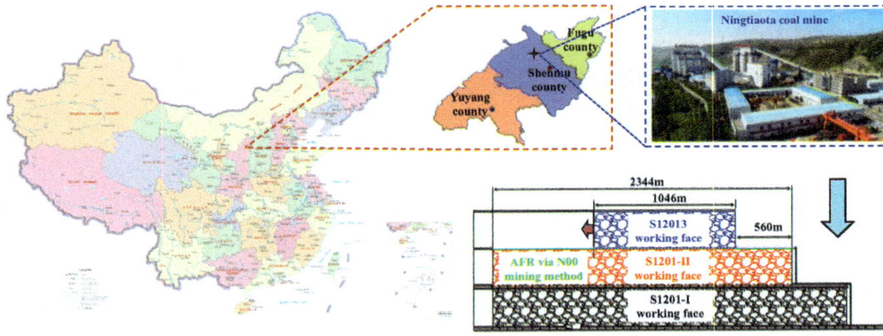

Fig. 4.1 General situation of the S1201-II working face

Table 4.1 Physical mechanical parameters of surrounding rocks

Lithology	γ/(kN·m^{-3})	E/GPa	μ	σ_c/MPa	σ_t/MPa	c/MPa	φ/(°)
Fine sandstone	23.1	11.38	0.23	22.25	0.64	1.27	40.91
Medium sandstone	22.8	10.16	0.19	19.70	0.58	1.12	41.33
Siltstone	23.9	17.58	0.19	24.90	0.86	1.87	38.29
Coal	12.6	7.59	0.18	13.35	0.36	0.85	39.69

the dip angle is about horizontal. The immediate floor is 16.3 m thick on average, the immediate roof is 2.2 m thick on average, the basic roof is 17.2 m thick on average, and the basic floor is 3.2 m thick on average. The physical and mechanical characteristics of the rock strata are displayed in Table 4.1.

4.1.2 Scheme Design

(1) **Overall scheme**

The roadway's cross-section size, surrounding rock support strength, and coal seam thickness are considered invariants in Type A–D schemes. The simulation is conducted with the pillar width, in-situ stress level, roof cutting height, and roof cutting angle. The specific scheme is as follows.

The scheme A refers to the comparison system that has varying pillar widths. The main comparison made by this kind of scheme is between the stress variation law and stress distribution of the surrounding rocks while gob-side entry driving with varying pillar widths and AFR without pillars. A_i, where $i = 1$–5, which has values of 3, 6, 9, 15, and 0 m, respectively.

The scheme B refers to the comparison system that has varying in-situ stresses. The main comparison made by this kind of scheme is between the surrounding rocks' stress distribution and variation law under various in-situ stress circumstances. There are two parts to this comparison scheme: AFR and gob-side entry driving. The degree of in-situ stresses is denoted by the variable B_j, where j ranges from 1 to 10. The values of $j = 1$–5 indicate 1.0, 1.5, 2.0, 2.5, and 3.0 of the in-situ stress level when gob-side entry driving mining method (A_1 scheme) is used; similarly, $j = 6$–10 indicates 1.0, 1.5, 2.0, 2.5, and 3.0 of the in-situ stress level when AFR mining method (A_5 scheme) is applied.

The scheme C refers to the comparison system that has varying roof cutting heights. The main comparison made by this kind of scheme is between the stress variation law and stress distribution of the surrounding rocks in the AFR. $C_{m,}$ where $m = 1$–5, which has values of 5, 7, 9, 11, and 13 m, respectively.

The scheme D refers to the comparison system that has varying roof cutting angles. The main comparison made by this kind of scheme is between the stress variation law and stress distribution of the surrounding rocks in the AFR. $D_{n,}$ where $n = 1$–5, which has values of 0°, 5°, 10°, 15°, and 20°, respectively.

(2) **Monitoring scheme**

The on-site geological conditions are taken into account while simulating a 400 m × 144 m × 180 m (width × height × thickness) range using $FLAC^{3D}$ numerical software. The model body's left and right limits are established horizontally, while the bottom boundary is fixed vertically, as shown in Fig. 4.2. Every rock strata in the model body is preserved in terms of both mechanical and physical parameters. A 90 m distance is chosen for the monitoring section from the model body's front surface.

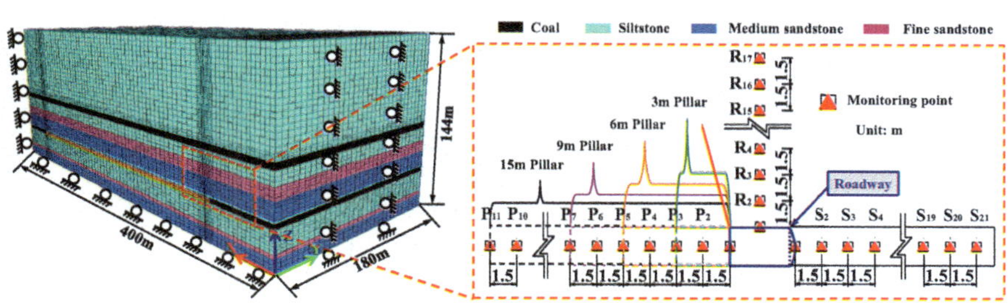

Fig. 4.2 Numerical monitoring scheme. Reprinted from Ref. Wang et al. (2021), Copyright © 2021 Elsevier Ltd.

4.1.3 Result Analysis

(1) Comparison analysis of different mine pillar width scheme

The pillar width is used as the variable for comparison analysis in the type A scheme. The side abutment pressure curve of the solid coal side and mine pillar is displayed in Fig. 4.3a. The variation curve of the roof stress under various pillar widths is displayed in Fig. 4.3b. The stress evaluating index diagram of the surrounding rocks under various pillar widths is displayed in Fig. 4.4.

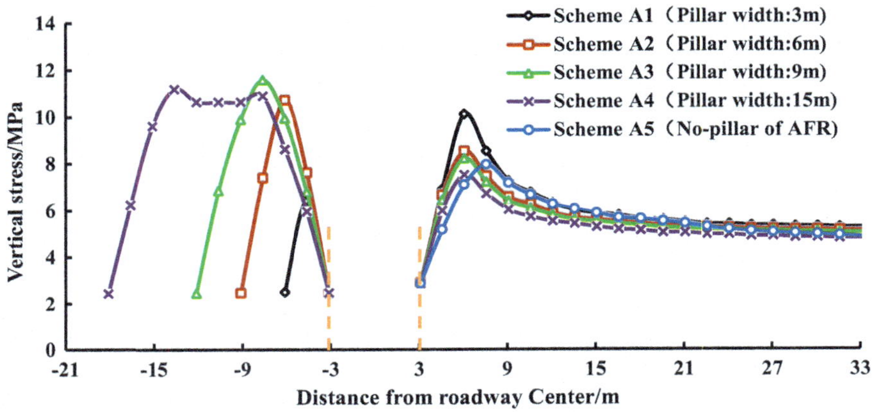

(a) The solid coal side and mine pillar side abutment pressure

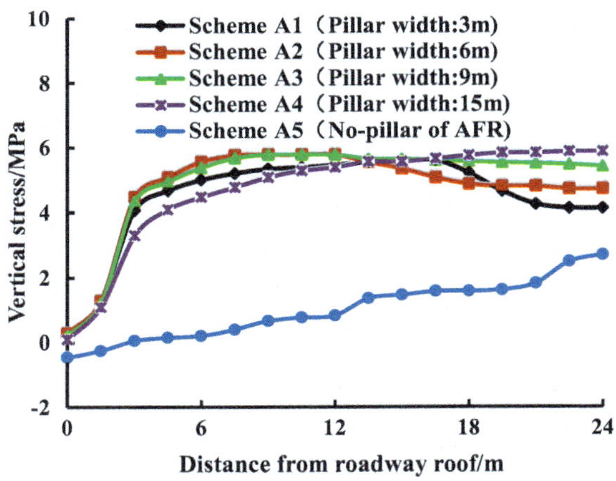

(b) The roof surrounding rock stress

Fig. 4.3 The surrounding rock stress curve of different pillar widths. Reprinted from Ref. Wang et al. (2021), Copyright © 2021 Elsevier Ltd.

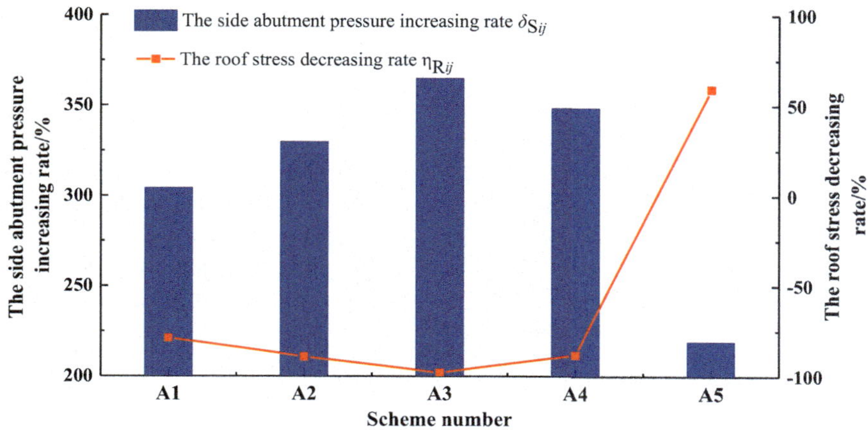

Fig. 4.4 The stress index of different pillar widths. Reprinted from Ref. Wang et al. (2021), Copyright © 2021 Elsevier Ltd.

By analyzing Figs. 4.3 and 4.4, the following conclusions may be drawn:

1) Analysis of the side abutment pressure

   (a) The stresses of the solid mine side and mine pillar side have an "asymmetric" characteristic as the pillar width rose, meaning that the stresses increased initially and subsequently reduced as they propagated away from roadway. The peak is on the solid mine side when the pillar width is 3 m; it moves to the mine pillar side when pillar width is 6–15 m.

   (b) Comparison of AFR schemes (A_1 –A_4) and conventional scheme (A_5): For schemes A1 through A5, the side abutment pressure peaks are 9.25, 8.76, 7.98, 7.85, and 7.68 MPa, in that order. Additionally, the matching side abutment pressure increase rates are 219.2%, 304.4%, 330.0%, 365.2%, and 348.8%, in that order. Compared to other pillar schemes, the AFR scheme (A_5) is substantially lower. Compared to the four pillar schemes, the AFR scheme's side abutment pressure peak is larger, with its peak located 4.5 m from the roadway, and the peak being roughly 3.0 m. According to the aforementioned results, roof cutting can lessen pressure stress peak and shift that far away from the roadway, which improves roadway stability.

2) Analysis of the roof stress

   (a) The conventional gob-side entry driving with mine pillar schemes is self-caparisoned (A_1 –A_4). The variation patterns in roof stress with the distance from the roadway roof are largely consistent throughout designs A_1 to A_4 with mine pillars, indicating a general pattern of "steep increase → slow increase → slow decrease." On the roadway roof, the vertical stress varies at the same location. The four schemes with mine pillars that corresponded to the decreasing rates of roof stress are − 78.4%, − 88.8%, − 97.6%, and − 88.0%, respectively.

(b) Comparison of AFR schemes (A_1–A_4) and conventional scheme (A_5): The AFR scheme (A_5) and the 3 m mine pillar scheme (A_1) are contrasted. For the two schemes, the average roof stresses are 1.02 MPa and 4.46 MPa, respectively. The AFR scheme's total roof stress is lower than the 3 m mine pillar scheme's, with a 77.1% reduction in comparison to the scheme's average roof stress. After a thorough comparison and study, it is possible to determine that the AFR design is more beneficial to the stability of the roadway roof because it has a more noticeable pressure-releasing effect than the 3 m mine pillar scheme.

(2) **Comparison analysis of in-situ stress level scheme**

The in-situ stress level serves as the variable for the comparative analysis in the Type B scheme. Figure 4.5 shows the variation curve of the side abutment pressure of the solid coal side in schemes B1–B5 of the gob-side entry driving and schemes B6–B10 of the AFR. The variation curve of the roadway roof stress in schemes B1–B5 and schemes B6–B10 is displayed in Fig. 4.6. Stress evaluation index diagram of surrounding rocks under varying in-situ stress levels is displayed in Fig. 4.7 for schemes B1–B5 and schemes B6–B10.

By analyzing Figs. 4.5, 4.6 and 4.7, the following conclusions may be drawn:

1) Analysis of the side abutment pressure

(a) Comparison of the traditional gob-side entry driving scheme with the AFR scheme in the part of side abutment pressure peak location: For schemes with different in-situ stress levels, the side abutment pressure of the AFR method (B_6–B_{10}) and the gob-side entry driving schemes (B_1–B_5) surge initially, dropped, and tended to stabilize as they traveled away from the roadway. For both mining techniques, the side abutment pressure peak positions are comparatively constant, and no discernible transfer under various in-situ stress conditions is seen. The conventional scheme and AFR mining method have pressure peaks that are roughly 3.0 m and 4.5 m from

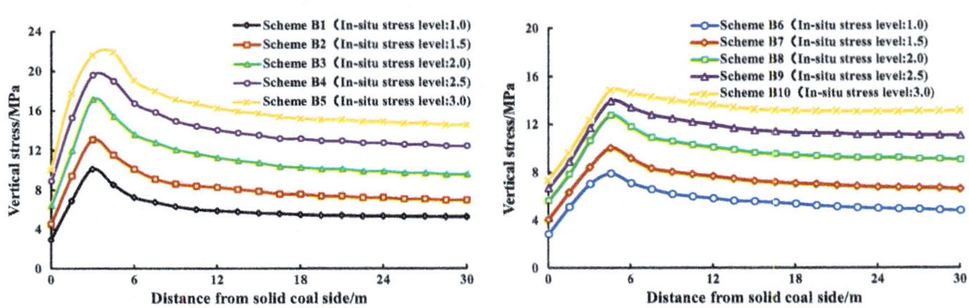

Fig. 4.5 The side abutment pressure curve of different in-situ stress levels. Reprinted from Ref. Wang et al. (2021), Copyright © 2021 Elsevier Ltd.

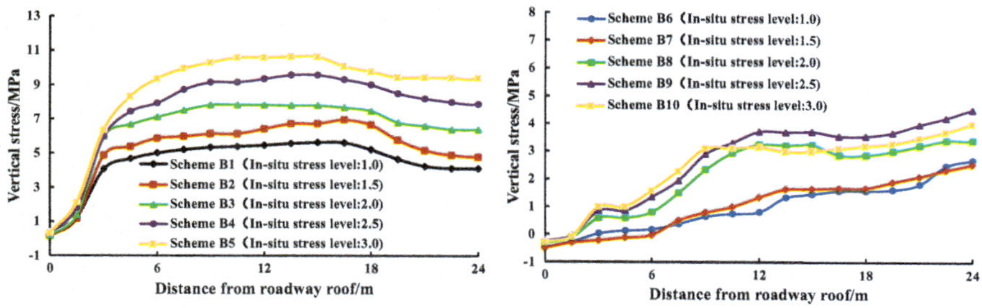

Fig. 4.6 The roof stress curve of different in-situ stress levels. Reprinted from Ref. Wang et al. (2021), Copyright © 2021 Elsevier Ltd.

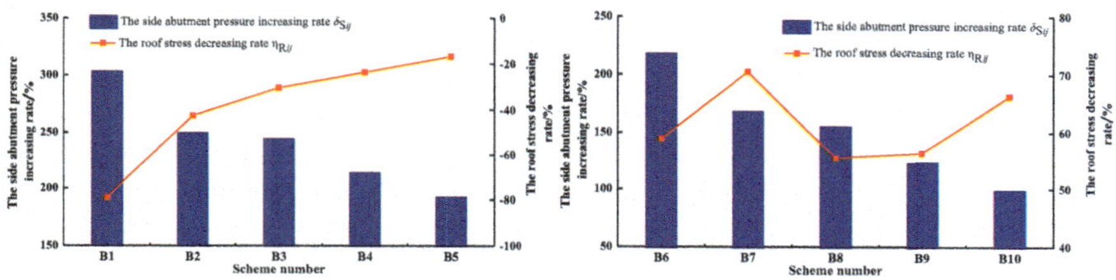

Fig. 4.7 The stress evaluation index of different in-situ stress levels. Reprinted from Ref. Wang et al. (2021), Copyright © 2021 Elsevier Ltd.

the roadway, respectively. The results show that roof cutting may move the side abutment pressure peak of the solid coal side to the deep section in both low and high in-situ stress schemes. This is beneficial for the roadway's stability.

(b) In terms of the side abutment pressure rising rate, contrasting the AFR technique with the traditional gob-side entry driving scheme: The AFR and gob-side entry driving techniques' side abutment pressure peaks grow to varying degrees as the in-situ stress increased. The AFR scheme's side abutment pressure growth rate decreases from 219.2% to 99.6%, resulting in a 119.6% relative decrease, while the in-situ stress level rises from 1.0 to 3.0. The conventional scheme's side abutment pressure increase rate drops from 304.4% to 193.5%, or a relative decrease of 110.9%. When mining activities moves to the deep section, both systems' side abutment pressure growth rates shows some degree of decline.

(c) Comparison of the side abutment pressure between the AFR scheme and the traditional gob-side entry driving with mine pillar scheme under high in-situ stress conditions: At a 3.0 in-situ stress level, the two methods display the largest peak values of the side abutment pressure. This in-situ stress level, for instance, is employed in a comparative study. In this scenario, conventional scheme's side abutment pressure peak is 22.01 MPa, 193.4% more than the initial rock stress. In contrast, the AFR scheme's is

14.94 MPa, which is 99.2% higher and substantially lower than the initial rock stress. The results show that at high in-situ stress conditions, the AFR scheme works better than the conventional scheme with mine pillar in terms of side abutment pressure.

2) Analysis of the roof stress

   (a) Comparison of the AFR mining method and the traditional gob-side entry driving method in the part of roof stress: As the in-situ stress level rises, the roof stresses of the AFR scheme and the gob-side entry driving scheme expand in different amounts. The former has a favorable rate of roof stress reduction, which varies between 55.8% and 77.0%. On the other hand, the latter shows a decreasing rate of negative roof stress, meaning that the average roof stress is larger than the average original rock stress. The findings show that the roof cutting successfully stops the stress from being transmitted between the AFR and the gob roof. As a result, it may successfully relieve the stress on the roof surrounding rock.
   (b) A comparison of the roof stress between the traditional gob-side entry driving with mine pillar scheme and the AFR scheme under high in-situ stress conditions: The gob-side entry driving scheme's average roof stress, at an in-situ stress level of 3.0, is 8.72 MPa, compared to 2.51 MPa for the AFR mining method. There is clear release of pressure from the former, which is 71.2% lower than the latter. The results show that, in high in-situ stress scenarios, the AFR mining method performs better in terms of roof stress than the gob-side entry driving method. As a result, the former is preferable for the roadway's stability.

(3) **Comparison analysis of roof cutting height scheme**

The Type C scheme uses the roof cutting height as the variable for the comparative analysis. Figure 4.8 shows the variation curve of the solid coal side's side abutment pressure in schemes C1–C5. Figure 4.9 shows the roadway roof stress fluctuation curve for schemes C1–C5. Figure 4.10 presents the stress assessment index diagram of the surrounding rocks at different roof cutting heights in schemes C1–C5.

After analyzing Figs. 4.8, 4.9 and 4.10, the following conclusions may be drawn:

1) Analysis of the side abutment pressure

   (a) Comparative study of side abutment pressures in different roof cutting height schemes: The pressure-releasing effect is strongly influenced by the height of the roof cutting. The stress peak of the roadway fell steadily as the roof cutting height climbed (from scheme C1 to C5); these values were 9.25, 8.76, 7.98, 7.85, and 7.68 MPa. The peak rose by 270.0%, 250.4%, 219.2%, 215.2%, and 207.2%, respectively, in comparison to the initial rock stress. This showed that the side abutment pressure concentration peak reduces with increasing roof cutting height, a trend that is more favourable to road stability.

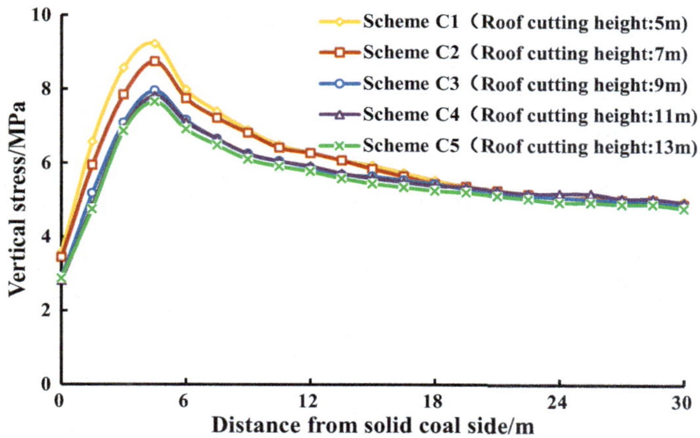

Fig. 4.8 The variation curve of the side abutment pressure under various roof cutting heights. Reprinted from Ref. Wang et al. (2021), Copyright © 2021 Elsevier Ltd.

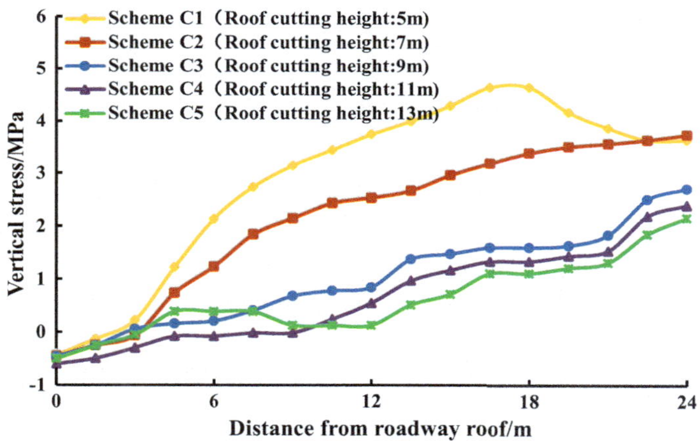

Fig. 4.9 The variation curve of the stress on the roadway roof at various heights of roof cutting. Reprinted from Ref. Wang et al. (2021), Copyright © 2021 Elsevier Ltd.

(b) Analysis of a reasonable roof cutting height: Variations in the side abutment pressure peak fluctuated by 1.9% at roof cutting heights of 9, 11, and 13 m. This suggests that after a certain height, further elevation of the roof cutting height did not significantly alter the side abutment pressure peak. Moreover, the expense and difficulty of building increased with the height of the roof cutting.

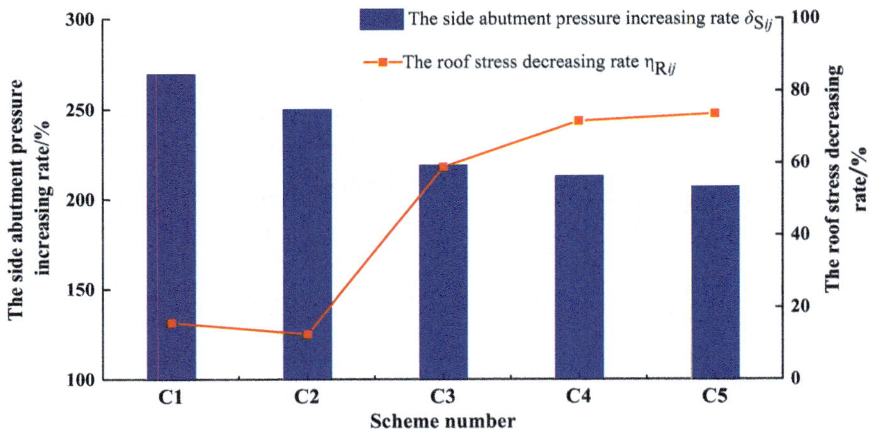

Fig. 4.10 The surrounding rocks' stress assessment index at various roof cutting heights. Reprinted from Ref. Wang et al. (2021), Copyright © 2021 Elsevier Ltd.

2) Analysis of the roof stress

Analysis of the roof stress in relation to various roof cutting height schemes: The total stress of the roadway roof steadily reduced as the roof cutting height grew (from scheme C1 to C5). The corresponding roof stress lowering rates were 16.0%, 12.8%, 59.2%, 72.0%, and 74.0%, respectively. The average roof stresses were 2.90, 2.18, 1.02, 0.70, and 0.65 MPa. The findings showed that the pressure-releasing effect became more noticeable and the roadway roof stress showed an overall decreasing trend with an increase in roof cutting height. Nevertheless, beyond a certain point, further elevation of the roof cutting height had no discernible impact on the roof stress.

(4) **Comparison analysis of roof cutting angle scheme**

The roof cutting angle is the variable used in the Type D scheme's comparison analysis. The variation curve of the side abutment pressure of the solid coal side in schemes D1–D5 is displayed in Fig. 4.11. The variation curve of the roadway roof stress in schemes D1–D5 is displayed in Fig. 4.12. The stress evaluating index diagram of the surrounding rocks at various roof cutting angles in schemes D1–D5 is displayed in Fig. 4.13.

After analyzing Figs. 4.11, 4.12 and 4.13, the following conclusions can be drawn:

1) Analysis of the side abutment pressure

The side abutment pressure peak of the solid coal side of the roadway declines first and subsequently climbs, measuring 8.66, 8.56, 7.98, 8.58, and 8.94 MPa, as roof cutting angle increases (from scheme D1 to D5). The side abutment pressure peak grows by 246.4%, 242.4%, 219.2%, 243.2%, and 257.6%, in that order, in relation to the initial rock stress. When the roof cutting angle is 10°, the side abutment

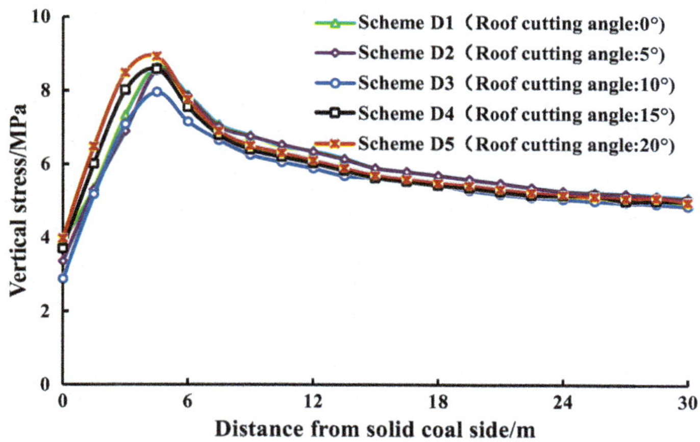

Fig. 4.11 The variation curve of the side abutment pressure under various roof cutting angles. Reprinted from Ref. Wang et al. (2021), Copyright © 2021 Elsevier Ltd.

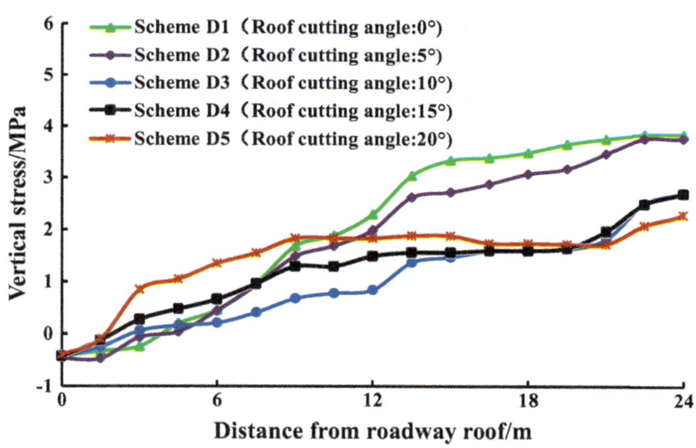

Fig. 4.12 The variation curve of the stress on the roadway roof under various roof cutting angles. Reprinted from Ref. Wang et al. (2021), Copyright © 2021 Elsevier Ltd.

pressure concentration peak is the smallest, 10.7% lower than in other designs, and the pressure-releasing action is the most evident. That suggests that this scheme's side abutment pressure promotes roadway stability more.

2) Analysis of the roof stress

(a) Analysis of the roof stress in relation to various roof cutting angle schemes: The overall roadway roof stress changes as the roof cutting angle increases (from scheme D1 to D5), with a drop followed by an increase. In that order, the decreasing rates of roof stress are 17.2%, 26.4%, 59.2%, 50.4%, and

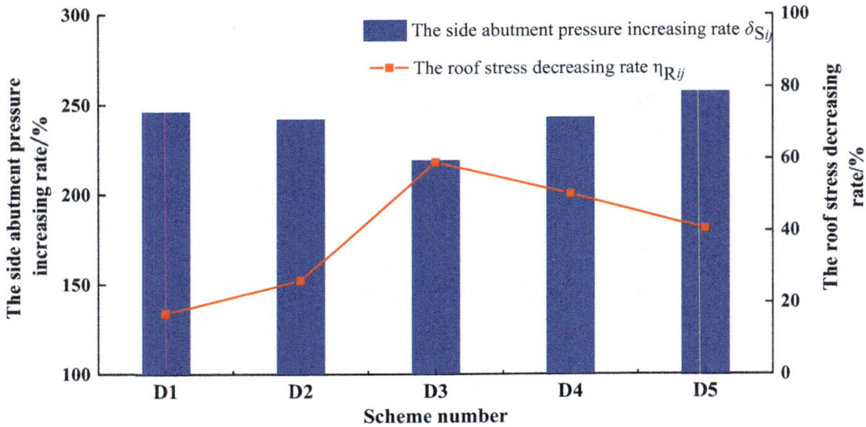

Fig. 4.13 The surrounding rocks' stress evaluating index under various roof cutting angles. Reprinted from Ref. Wang et al. (2021), Copyright © 2021 Elsevier Ltd.

40.8%. The average stresses on the roof are 1.84, 1.02, 1.24, 1.48, and 2.07 MPa. While the range of roof cutting angles from 10° to 15° shows a significant improvement over the other three designs, the values and trends of roof stress differs slightly across them.

(b) Evaluation of a suitable angle for roof cutting: The roof cutting angle was within a reasonable range. A suitable roof cutting angle can effectively and totally halt the stress transfer between the gob roof and the AFR. Moreover, it may facilitate the gob's underlying strata collapsing and subsiding while lowering friction resistance, all of which are advantageous for the development of the roadway and the management of the surrounding rock strata.

4.2 The AFR with Deep Buried Depth

Mine pressure is the main cause of the roadway control problems. As a new mining method, AFR can release the stress effectively, which widely are adopted in shallow mines. However, the adaptability in deep roadway with high stress roof breakage still needs further study. This section takes 2215 working face of Suncun Mine as the research object, and carries out two kinds of numerical tests, namely, the comparison between deep high stress gob-side entry driving and AFR.

4.2.1 Research Object

The Suncun Mine located in Xinwen mine area of China, with a maximum mining depth of 1501 m and a horizontal in-situ stress of 37.9 MPa, which is the deepest mine in China, as shown in Fig. 4.14. The physical and mechanical characteristics of the rock strata are displayed in Table 4.2. The average coal seam thickness of the 2# coal on the 2215 working face in the No. 2 mining area is 2.5 m. The average thickness of the siltstone immediate roof is approximately 2.5 m, the average thickness of the sandstone basic roof is approximately 7 m, the average thickness of the fine sandstone immediate floor is approximately 3 m, and the average thickness of the siltstone main floor is approximately 3 m. In mining area 2, the gob-side entry driving method excavating the roadway by leaving tiny mine pillars along the gob is used. The roadway height is 2.5 m, its width is 3.8 m, and the mine pillar's breadth is 5 m. Bolts (Φ20 mm × 2800 mm) and anchor cables (Φ22 mm × 5000 mm) support the roof, while bolts (Φ20 mm × 2000 mm) support the two sides. The spacing of anchor cables and bolts are 1000 mm × 800 mm.

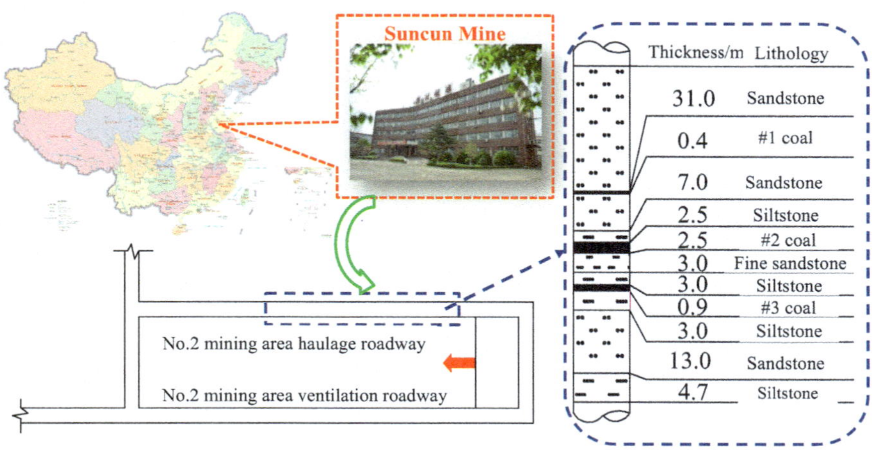

Fig. 4.14 Geographical location and working face situation

Table 4.2 Mechanical parameters of rock mass

Lithology	Bulk modulus/GPa	Shear modulus/GPa	Tensile strength/MPa	Cohesion/MPa	Internal friction angle/(°)
Sandstone	3.03	1.12	3.5	2.4	36
Fine sandstone	1.5	1.04	3.2	2.2	34
Siltstone	3.07	2.03	3.8	3.7	40
Coal	0.86	0.46	2.1	1.4	30

Fig. 4.15 Deformation and failure of the roadway

The surrounding rock of the roadway with the typical mine pillar excavation and the traditional support method is severely broken as a result of the high ground stress and excavation disturbance. The deformation is large, the net and bolts (anchor cables) are broken seriously, and roof falling accidents often occur, as shown in Fig. 4.15.

The typical section is chosen to monitor the convergence deformation of surrounding rock, the load on the support members, and the roof's breaking range in order to provide light on the deformation and failure process of the roadway. Figure 4.16 displays the monitoring data for the roof and floor convergence deformation, the convergence deformation of the two sides, and the bolt (cable) stress. The center, left, and right sides of the roof have a 6 m borehole dug into them. A borehole detector detects the rock surrounding the hole. The surrounding rock of the roof is separated into three zones based on the findings of the detection: the original rock strength zone, the strength recovery zone, and the strength degradation zone. As seen in Fig. 4.17, the black fill in the borehole represents the fracture distribution region, while the white fill represents the integrity region.

It can be seen from the analysis of Figs. 4.16 and 4.17.

(1) From the period of roadway excavation to the period of coal mining, the convergence deformation of the roadway continues to increase. The surface displacement of roadway increases rapidly within 30 days of excavation. The growth rate of roadway surface displacement slows down on the 30–60th day. The deformation is characterized by "the convergence deformation between the two sides > the convergence of the roof and floor," with the convergence deformation of the two sides reaching 57.2 cm and the convergence deformation of the roof and floor reaching 62.1 cm.
(2) Following installation, the bolt's force first rises, then falls, and ultimately tends to stabilize. Within 15 days after installation, the force of the bolt increases

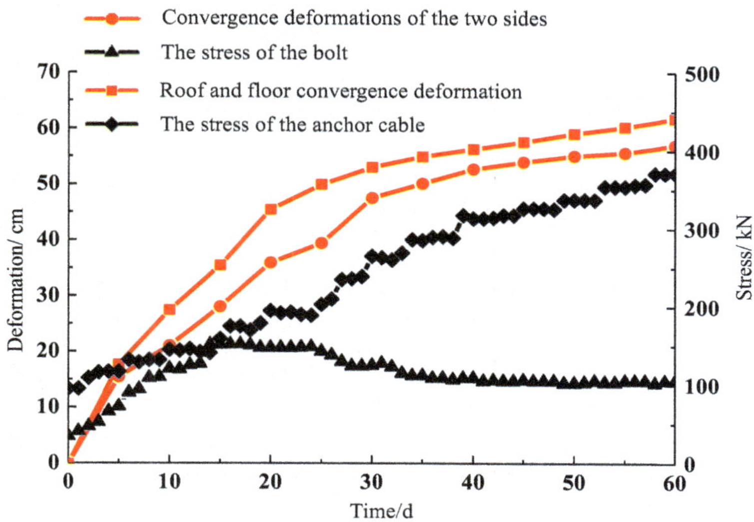

Fig. 4.16 Convergence deformation of typical section and stress of supporting members

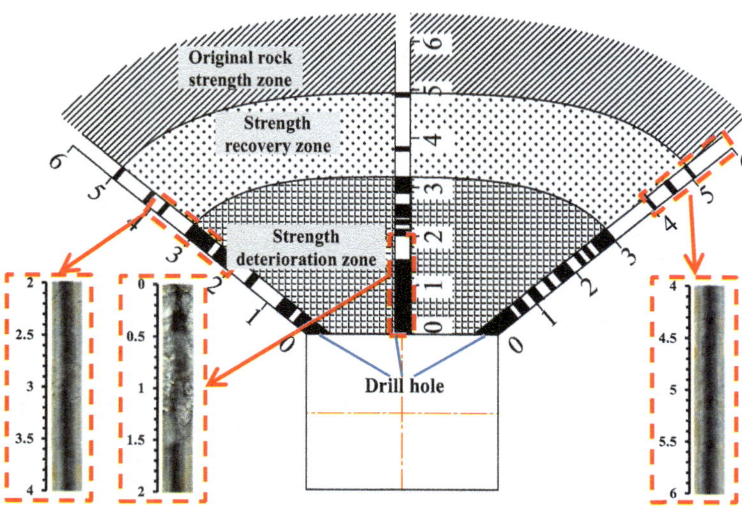

Fig. 4.17 Roof division of the roadway (unit: m)

rapidly, reaching a maximum of 152.2 kN. From the 15th to the 25th day, the force of the bolt begins to decrease. After 33 days, the force of the bolt is basically unchanged, about 103.5 kN. On the 60th day, the cable stress reaches 371.1 kN, the ultimate breaking force is 411 kN, and the strength utilization rate reaches 90.3%.

(3) The overall loosening of roadway roof is large, and the strength of surrounding rock is weakened obviously. Among them, the strength deterioration zone ranges from 0 to 3.1 m, surrounding rock is very broken, and the strength deterioration is serious. The strength recovery zone ranges from 3.1 to 4.9 m, and the rock mass is relatively intact with a few micro-cracks. The original rock strength zone is 4.9 m away, and the surrounding rock is intact without obvious cracks.

The following monitoring findings are used to assess the primary causes of roadways deformation and support failure. On the one hand, the roadway is situated in an area with significant ground stress because of the impact of enormous buried depth and excavation disturbance. High stress cannot be efficiently released by the gob-side entry driving method. Instead, the stress is concentrated at the mine pillar, increasing the mine pillar's deformation and eventual disintegration. Conversely, the roadway uses anchor cables and bolts for stability, with the majority of the bolts in a weakening region. It is difficult to play the supporting role, the anchor cable force is too large, close to the ultimate breaking force, and the safety reserve is low. Under this support condition, the range of surrounding rock loosening and deterioration is further expanded. The severity of the roadway surrounding rock deformation problem is increasing, making roof control increasingly challenging.

4.2.2 Scheme Design

(1) **Overall scheme**

The comparison schemes of gob-side entry driving and AFR include scheme A (different mine pillar width) and scheme B (different ground stress). The AFR scheme includes scheme C (different roof surrounding rock strength) and scheme D (different roof cutting height). Scheme A–D treats the width of the mine pillar, the magnitude of the ground stress, the strength of the roof surrounding rock, and the roof cutting height as variables in turn, while treating the roadway section size, the original bolting and metal mesh support as invariants. The following are the particular schemes.

Scheme A (different mine pillar width comparison scheme): This kind of scheme mainly compares the deformation and stress changes of roadway surrounding rock. The numbers are A_i, where i = 1 to 5, corresponding to the widths of mine pillars of 3 m, 5 m, 8 m, 15 m and 0 m respectively.

Scheme B (different ground stress comparison schemes): This kind of scheme mainly compares the deformation and stress changes of surrounding rock under different ground stress conditions, taking the ground stress as the variable. The numbers are B_j, where j = 1 to 10. Where j = 1–5 corresponds to the ground stress levels of 0.75, 1.0, 1.25, 1.5 and 1.75 respectively. j = 6–10 corresponds to the ground stress levels of 0.75, 1.0, 1.25, 1.5, 1.75 under the condition of AFR (scheme A5).

Scheme C (different roof surrounding rock strength comparison scheme): This kind of scheme mainly simulates the reinforcement effect of grouting on the surrounding rock, and simulates the improvement of surrounding rock strength by increasing their strength coefficient. The values C_m, where m = 1–5, correspond to the strength coefficient of the surrounding rock of the roadway roof being 1.0, 1.1, 1.2, 1.3, and 1.4, respectively, when the strength of the rock is taken as a variable.

Scheme D (different roof cutting height comparison scheme): This kind of scheme is mainly based on the simulation analysis of the core parameter—cutting height of AFR. Using the roof cutting height as a variable, the numbers are D_n, where n = 1–5, corresponding to roof cutting heights of 4 m, 6 m, 8 m, 10 m, and 12 m, respectively.

(2) **Monitoring scheme**

Based on the geological conditions on site, numerical calculations are conducted using FLAC^{3D} numerical software, with a model size of 280 m (width) × 70 m (height) × 70 m (thickness). The bottom boundary of the model body is fixed vertically, and the left and right boundaries are fixed horizontally. The stress in the X and Y directions is 23.4 MPa, and the stress in the Z direction is 19.4 MPa. The physical and mechanical parameters of each rock strata in the model body are consistent with those on site. The model body is shown in Fig. 4.18.

Set up a monitoring section at a distance of 35 m from the front surface of the model, where the monitoring points for the roof of the roadway and the solid coal side are numbered R_{i-j} and S_{i-j} (i = 1 or 2, respectively, representing the gob-side entry driving and the AFR. j represents the monitoring point number). The layout and position of each monitoring point are shown Figs. 4.18 and 4.19.

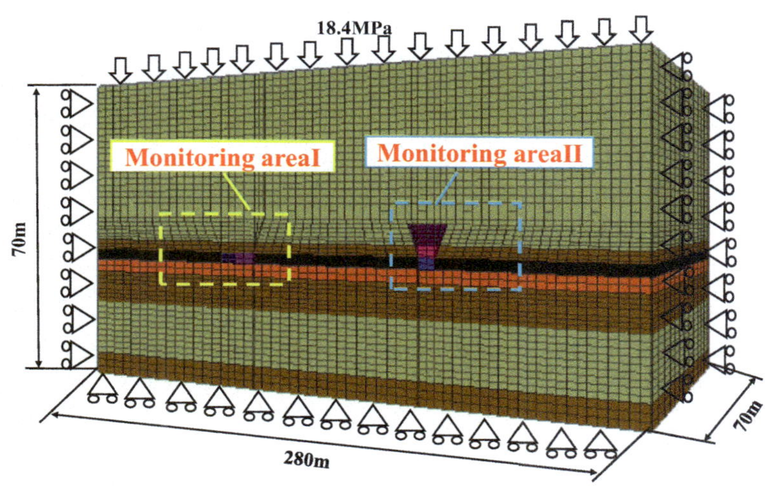

Fig. 4.18 Diagram of model body and boundary conditions

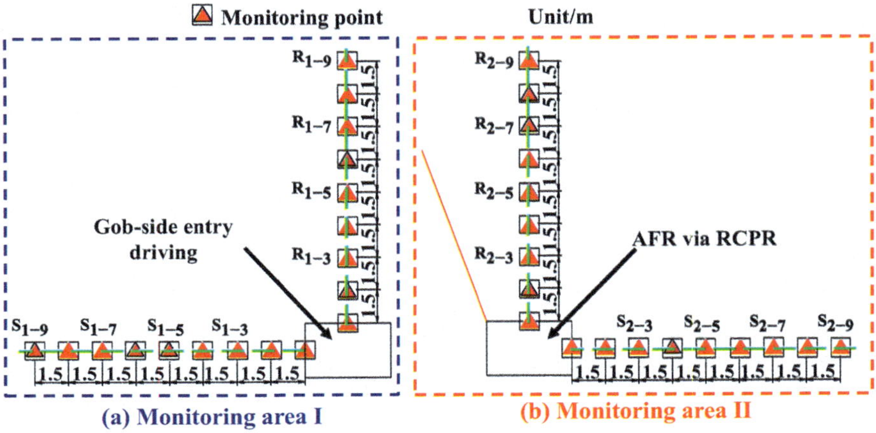

Fig. 4.19 The monitoring scheme of numerical simulation

4.2.3 *Result Analysis*

To better analyze stress changes and control mechanisms of surrounding rocks in different schemes of roadways, quantitative evaluation indexes such as side abutment pressure increase rate $(\delta_{S_{ij}})$, roof stress release rate $(\delta_{R_{ij-k}})$, and surrounding rock deformation control rate $(\eta_{D_{ij}})$ are established.

1) Side abutment pressure rising rate $(\delta_{S_{ij}})$, the percentage increase in side abutment pressure between the mine pillar and solid coal seam after stable roadway excavation compared to the original rock stress.

$$\delta_{S_{ij}} = \frac{S_{ij-max} - S_{ij-ini}}{S_{ij-ini}} \times 100\% \tag{4.1}$$

where

$\delta_{S_{ij}}$—Side abutment pressure increase rate of roadway in *ij* scheme. It shows how the side abutment pressure of the roadway in the *ij* scheme has increased in comparison to the initial rock stress. Where, *i* represents different scheme numbers, *i* = A–D, respectively represents the four types of schemes designed. The *j* represents the subschemes of different schemes, *j* = 1–5, among which Scheme B, *j* = 1–10.

$S_{ij\text{ - }max}$—Peak stress of roadway in *ij* scheme, MPa.

$S_{ij\text{ - }ini}$—Initial rock stress of roadway in *ij* scheme, MPa.

2) Roof stress releasing rate $(\delta_{R_{ij-k}})$, the percentage reduction of roof stress from original rock stress after roadway excavation stability.

$$\delta_{R_{ij-k}} = \frac{S_{ij-ini} - R_{ij-k}}{S_{ij-ini}} \times 100\% \tag{4.2}$$

where:

$\delta_{R_{ij-k}}$—In *ij* scheme, roof stress release rate of roadway at No. *k* measuring point. It indicates the reduction degree of roof stress at the No. k measuring point in the roadway roof relative to the original rock stress in *ij* scheme. Among them, the meanings of *i* and *j* are the same as Eq. (4.1).

R_{ij-k}—Roof stress at the No. *k* measuring point of roadway roof in scheme *ij*, MPa.

3) Surrounding rock deformation control rate $(\eta_{D_{ij}})$, percentage reduction of roof deformation after roadway excavation stability compared with the maximum roof deformation of similar schemes. The calculation formula is as follows:

$$\eta_{D_{ij}} = \frac{D_{imax} - D_{ij}}{D_{imax}} \times 100\% \tag{4.3}$$

where

$\eta_{D_{ij}}$—Deformation control rate of roadway surrounding rock in *ij* scheme. It represents the improvement degree of roadway surrounding rock deformation control compared with the subscheme with largest roadway roof deformation in scheme i. The higher the $\eta_{D_{ij}}$, the more obvious the control effect of roadway surrounding rock is. Among them, the meanings of *i* and *j* are the same as Eq. (4.1).

D_{ij}—The deformation of roadway roof in *ij* scheme, mm.

D_{imax}—Maximum roof deformation of all subschemes in scheme i, mm.

(1) **Comparison analysis of different mine pillar width scheme**

This type of scheme (A_1–A_5) is compared and analyzed using the width of the mine pillar as a variable. Side abutment pressure with different mine pillar width is shown in Fig. 4.20. Roadway roof stress variation curves are shown in Fig. 4.21. In order to comprehensively compare and analyze the stress changes and deformation control laws of the surrounding rock of each scheme's roadway, various evaluation indexes are summarized, as shown in Fig. 4.22. In Fig. 4.22, the stress release rate of the roof is selected as the typical representative points for the shallow and deep parts of the roadway roof at measurement points 3 and 7.

By analyzing Figs. 4.20, 4.21 and 4.22, it can be seen that:

1) **Analysis of the side abutment pressure**

   (i) As the width of the mine pillar increases, the overall side abutment pressure shows an "asymmetric" feature of first increasing and then decreasing, during the development of the mine pillar on the left side and the solid coal side on the right side of the roadway towards the deep. When the width of the mine pillar is 3–5 m, the peak side abutment pressure is located on the solid coal side. When the width of the mine pillar is 8–15 m, the peak side abutment pressure shifts to the side of the mine pillar.

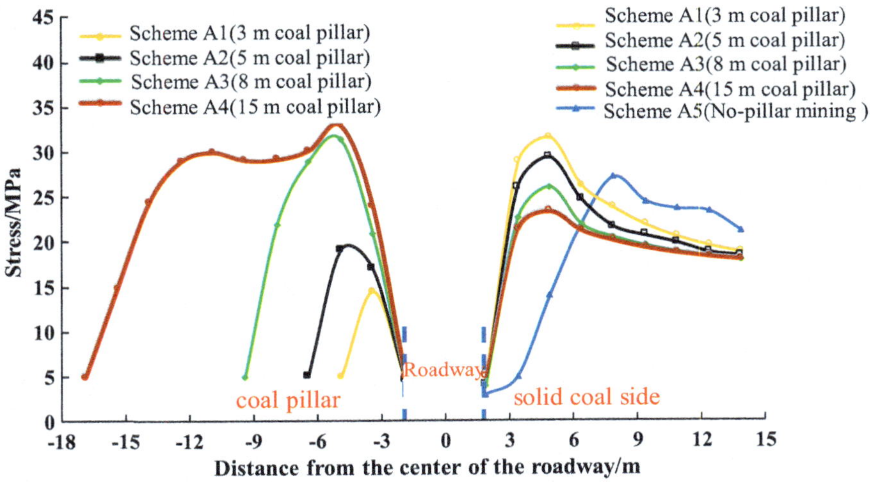

Fig. 4.20 Side abutment pressure with different mine pillar width

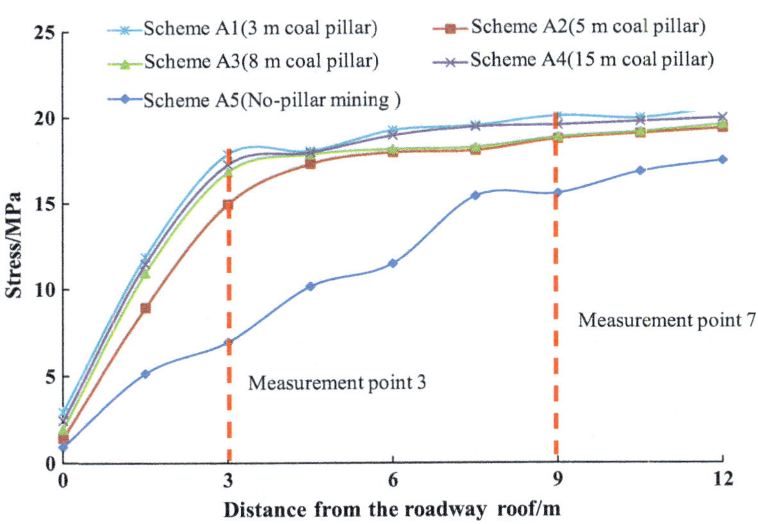

Fig. 4.21 Variation of roof stress with different mine pillar width

(ii) The comparison and analysis stress peak of the AFR and conventional method are as follows. In these schemes (scheme A1–A4), the average of side abutment pressure rising rate of the solid coal side is 58.0%, with a maximum peak value of 33 MPa. For the solid coal side, the AFR scheme (scheme A5) has a side abutment pressure rising rate of 36.2%, peaking at

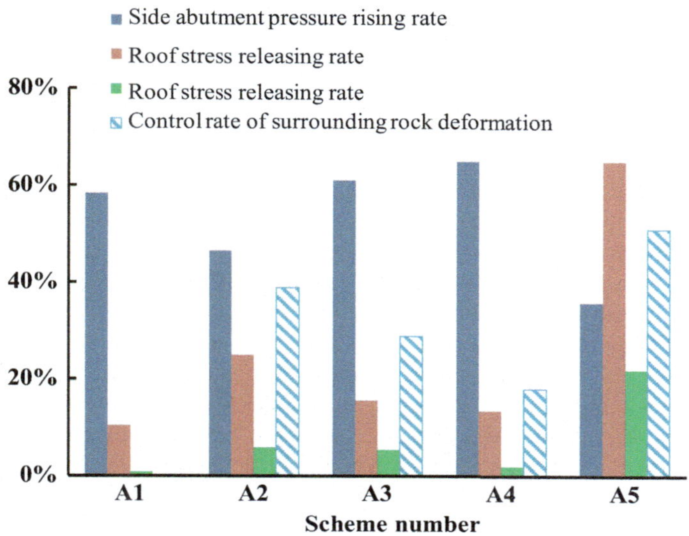

Fig. 4.22 Quantitative evaluation indexes of different mine pillar width

27.2 MPa. Meanwhile, the latter is 6 m further away from the surface of the roadway than the former in the part of pressure peak. The aforementioned findings show that roof cutting pressure release is efficient ways to lower the solid coal side's peak side abutment pressure.

2) **Analysis of the roof stress releasing effect**

   (i) The farther the roadway roof is from the roadway surface in A_1–A_5 schemes, the greater the stress on the roof and the closer the stress changes are to stability. The roof stress (scheme A1–A4) tends to basically approach original rock stress as it is about 6 m away from the surface of the roadway. The roof stress of the AFR by roof cutting scheme (scheme A5) still does not reach the original rock stress at a distance of 12 m from the roadway surface. This indicates that roof cutting pressure releasing can create a larger pressure releasing zone above the roadway roof, making it easier to control the surrounding rock of the roadway roof.

   (ii) The average pressure releasing rate of the roof (scheme A1–A4) and the pressure releasing rate of the roof in the AFR by roof cutting scheme (scheme A5) are 16.2% and 65.2% at the shallow measurement point 3, and 3.6% and 22.3% at the deep measurement point 7, respectively. The pressure releasing rate of the same measurement point in the latter is significantly higher than that of the former. This indicates that the stress on the roof is effectively released by roof cutting. The closer it is to the surface of the roadway, the more obvious the pressure releasing effect on the roof is.

3) **Analysis of the surrounding rock deformation control**

(i) The deformation control characteristics of the surrounding rock in the gob-side entry driving scheme are as followed. When the width of the mine pillar is 3–15 m, the deformation control rates of the surrounding rock are 0%, 39.3%, 29.1%, and 18.5%, respectively, presenting the characteristics of "5 m mine pillar scheme > 8 m mine pillar scheme > 15 m mine pillar scheme > 3 m mine pillar scheme".

(ii) Comparative analysis of the deformation control effect of the surrounding rock between the AFR by roof cutting and gob-side entry driving are as followed. The average deformation control rate (scheme A1–A4) and the AFR by roof cutting scheme (scheme A5) are 21.7% and 51.7%, respectively, with the latter being significantly higher than the former. This indicates that the roof cutting can cut off the stress transfer between the gob and the roadway, causing the roadway to be in a pressure releasing zone and effectively controlling roadway deformation.

According to A scheme, following conclusions could be obtained as follow. Compared to the gob-side entry driving, the AFR by roof cutting forms a larger pressure releasing zone. It can reduce the side abutment pressure of the solid coal side, make the peak value shift to the deep. The surrounding rock of the roadway roof is easier to control, and the solid coal side is more stable.

(2) **Comparison analysis of different ground stress level scheme**

The B_1–B_{10} scheme is compared and analyzed by using the ground stress as a variable. The side abutment pressure variation curves of the solid coal side in the gob-side entry driving scheme (scheme B1–B5) and the AFR by roof cutting scheme (scheme B6–B10) are shown in Fig. 4.23a, b, respectively. The stress variation curves of the roadway roof in the gob-side entry driving scheme and the AFR by roof cutting scheme are shown in Fig. 4.24a, b, respectively. In order to comprehensively compare and analyze the stress changes and deformation control laws of the surrounding rock of each scheme, various evaluation indexes are summarized, as shown in Fig. 4.25. In Fig. 4.25, the pressure releasing rate of the roof is selected from the measurement points 3 and 7 of the roadway roof as typical representative points for shallow and deep roof evaluation.

By analyzing Figs. 4.23, 4.24 and 4.25, it can be seen that:

1) **Analysis of the side abutment pressure**

(i) The side abutment pressure peak characteristics are analyzed as follows. In every scheme, the side abutment pressure peak in solid coal side rises with increasing ground stress, while the solid coal side's surface is comparatively stable from the peak. The solid coal side of gob-side entry driving and AFR by roof cutting have side abutment pressure increasing rates of 21.2% and 10.3%, respectively, and are 3 and 6 m from the surface, respectively, when the ground stress level rises to 1.75. This shows that in a high stress setting, the side abutment pressure peak of the AFR solid coal side is located farther

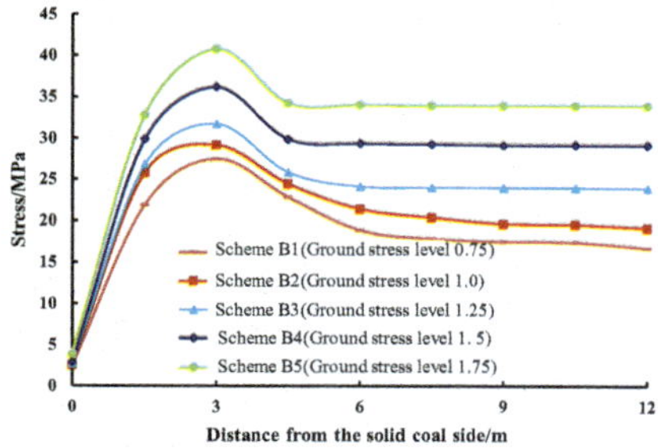

(a) Gob-side entry driving scheme

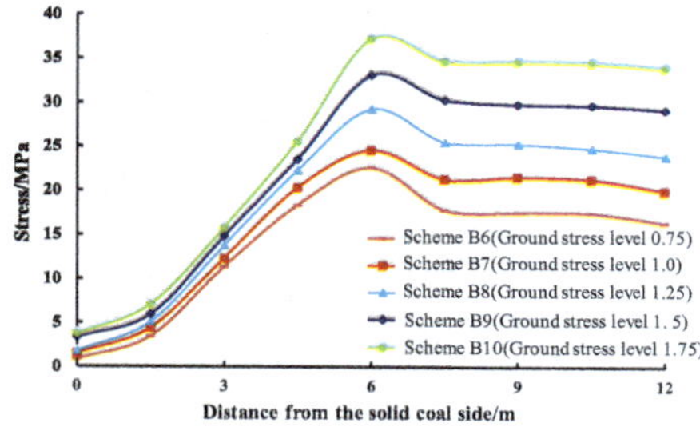

(b) The AFR scheme

Fig. 4.23 Variation of side abutment pressure with different ground stress level

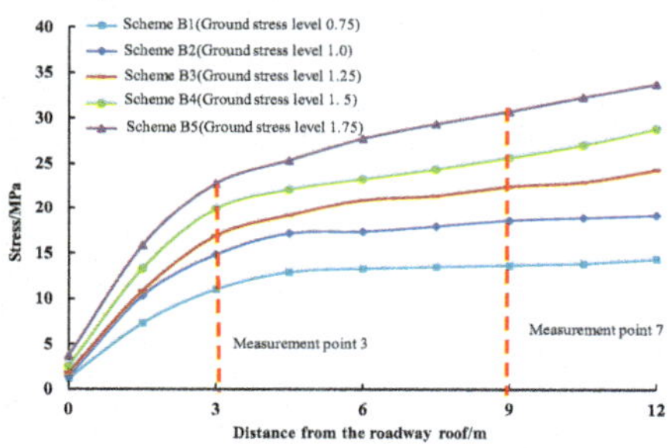

(a) Gob-side entry driving scheme

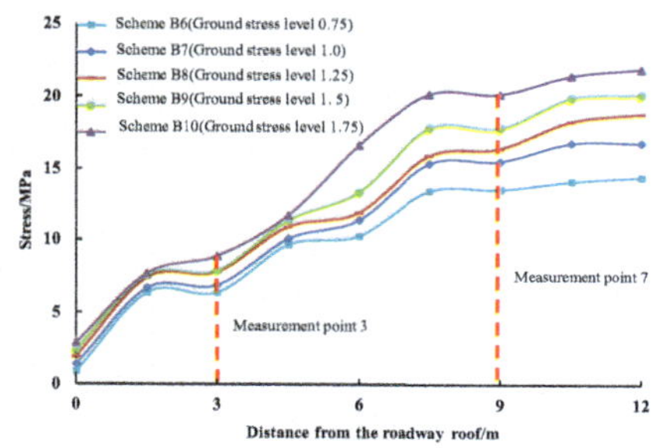

(b) The AFR scheme

Fig. 4.24 Variation of roof stress with different ground stress level

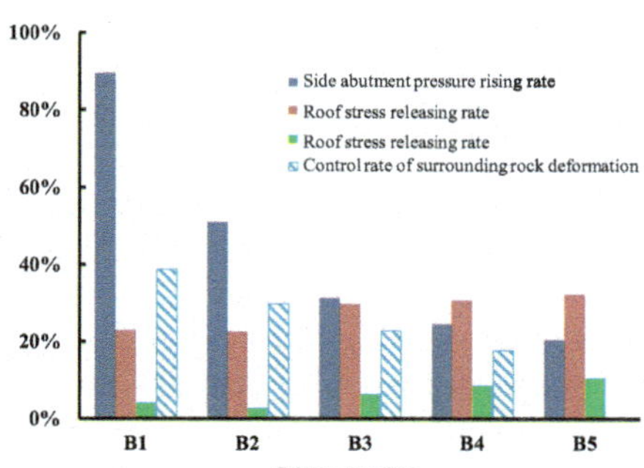

(a) Gob-side entry driving scheme

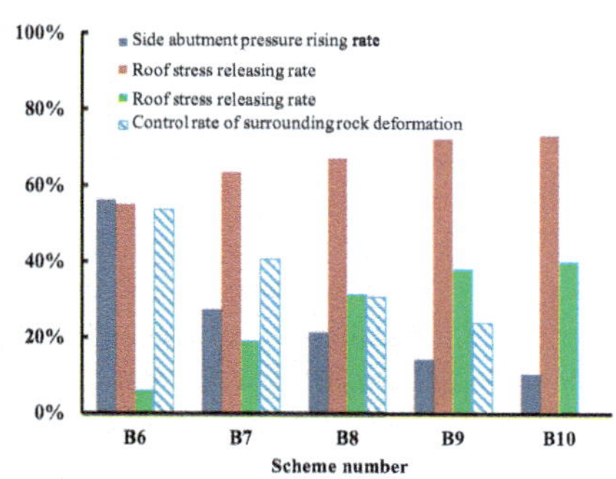

(b) The AFR scheme

Fig. 4.25 Quantitative evaluation indexes

away from the surface of the roadway solid coal side than the side abutment pressure peak of the roadway solid coal side of gob-side entry driving.

(ii) As the ground stress increases, the side abutment pressure rising rate on the solid coal side of AFR produced by roof cutting gradually decreases. Raising the ground stress level from 0.75 to 1.75 causes the solid coal side's side abutment pressure to rise more slowly, from 56.3% to 10.3%. This demonstrates that as ground stress increases, the side abutment pressure peak of the AFR solid coal side decreases with respect to the initial rock stress.

2) **Analysis of the roof pressure releasing effect**

(i) With the increase of ground stress, the stress on the roof with two mining methods increases. When the ground stress level increases to 1.75, the pressure releasing rates of the roof of AFR and the gob-side entry driving are 73.5% and 33.8% at the shallow measurement point 3, and 40.3% and 11.7% at the deep measurement point 7, respectively. The above results indicate that the stress releasing rate of the AFR at the same position of the roof is significantly lower than that of the gob-side entry driving, with better roof pressure releasing effect and more stable surrounding rock.

(ii) The analysis of roof pressure releasing degree of the AFR by roof cutting is as follows. As the ground stress increases, the stress releasing rate of the AFR roof gradually increases. In the B_6–B_{10} schemes of AFR by roof cutting, the stress releasing rate of the roof increases from 55.1% to 73.5% at the shallow measurement point 3, and from 7.8% to 40.3% at the deep measurement point 7. This indicates that the greater the ground stress, the higher the degree of stress release on the roof of the AFR by roof cutting, and the more obvious the pressure releasing.

3) **Analysis of the surrounding rock deformation control**

As the ground stress increases, the deformation control rate of the surrounding rock of the AFR by roof cutting gradually decreases. The deformation control rates of surrounding rock in B_6–B_{10} schemes of the AFR by roof cutting are 54.2%, 41.6%, 31.7%, 24.4%, and 0%, respectively. This indicates that in high stress environments, it is difficult to control the surrounding rock of the roadway. To ensure the stability of the roadway surrounding rock, it is necessary to improve the support strength of the roadway or the bearing capacity of the surrounding rock itself while roof cutting to relieve pressure.

Based on the comprehensive analysis of B scheme, the following conclusions can be drawn. The peak side abutment pressure and roof stress of the solid coal side of the AFR by roof cutting are lower than those of the gob-side entry driving under varying ground stress circumstances.

(3) **Comparison analysis of different surrounding rock strength scheme**

The C_1–C_5 scheme is compared and analyzed by using the width of roof surrounding rock as a variable. Figure 4.26 displays the side abutment pressure variation curves

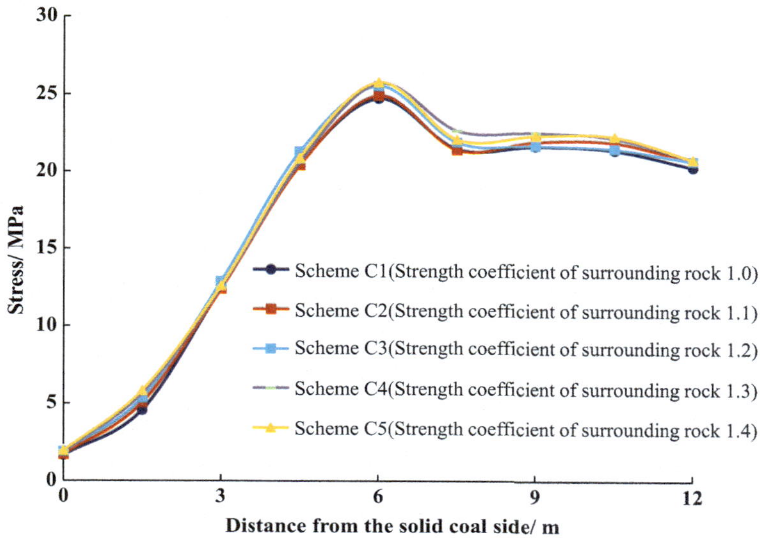

Fig. 4.26 Variation of side abutment pressure

for the mine pillar and solid coal side in each design, whereas Fig. 4.27 displays the roadway roof's stress variation curve. Various evaluation indexes were compiled, as shown in Fig. 4.28, so as to thoroughly evaluate and assess the surrounding rock stress changes and deformation control laws of various schemes. In Fig. 4.28, the pressure release rate of the roof was selected from the measurement points 3 and 7 of the roadway roof as typical representative points for shallow and deep roof evaluation.

By analyzing Figs. 4.26, 4.27 and 4.28, it can be seen that:

1) **Analysis of the side abutment pressure**

When the strength coefficient of the roof surrounding rock is 1.0–1.4, the side abutment pressure of the solid coal side in the AFR increased at rates of 24.2%, 25.4%, 28.1%, 29.2%, and 29.3% in the AFR by roof cutting scheme (scheme C1–C5). In addition, each scheme's solid coal side's peak side abutment pressure is located around 6 m below the surface of the solid coal side. This indicates that increasing the strength of the surrounding rock within a certain range of the roadway roof has a relatively small impact on the changes in the side abutment pressure of the solid coal side.

2) **Analysis of the roof pressure releasing effect**

At the shallow measuring point 3 of the AFR by roof cutting, the maximum pressure releasing rate of the roof is 65.1%, and the lowest is 57.8%, when the strength coefficient of the surrounding rock of the roof is between 1.0 and 1.4. The maximum pressure releasing rate of the roof at measuring point 7 in the depth of the roof is

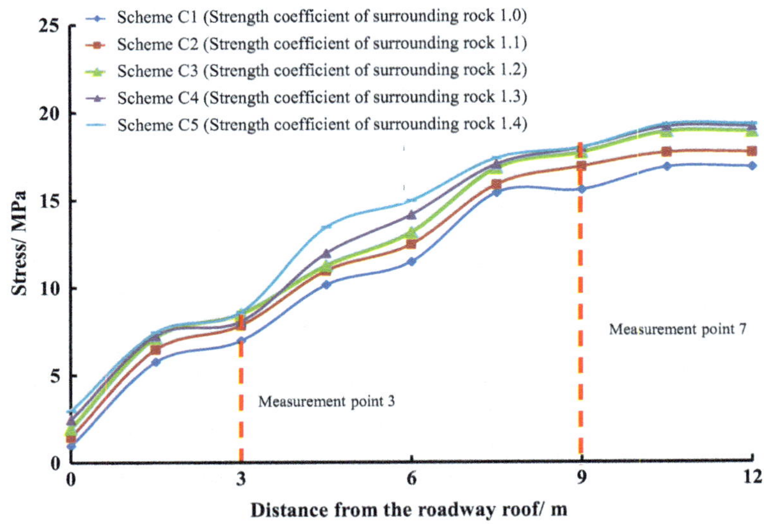

Fig. 4.27 Variation of roof stress

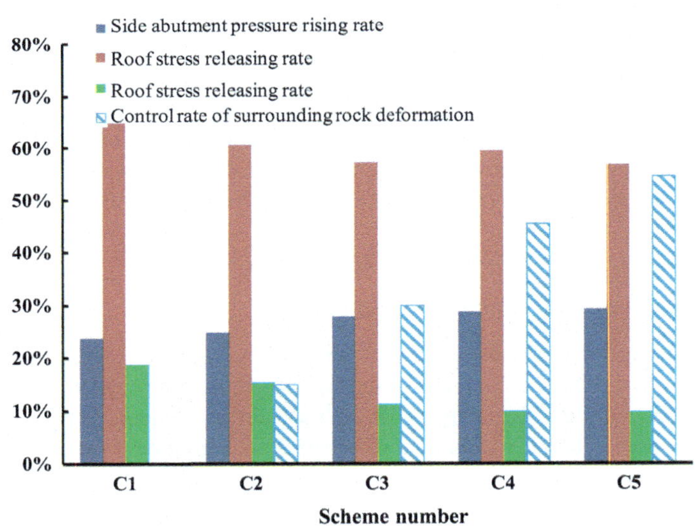

Fig. 4.28 Quantitative evaluation indexes of different surrounding rock strength

18.2%, and the minimum is 10.9%. The pressure releasing effect of roadway roof is not significantly different. This indicates that increasing the strength of surrounding rock within a certain range of the roadway roof has a relatively small impact on the stress changes of the roof surrounding rock.

3) **Analysis of the surrounding rock deformation control**

When the strength coefficient of the roof surrounding rock increases from 1.0 to 1.4, the deformation control rate of the AFR surrounding rock increases from 0 to 55.6%, and the surrounding rock control effect becomes better and better. This indicates that increasing the strength of surrounding rock within a certain range of the roadway roof can effectively control the deformation of the roadway surrounding rock and maintain the stability of the roadway.

Based on the comprehensive analysis of scheme C, the following conclusions can be drawn. Improving the strength of the surrounding rock of the roadway roof can effectively control the deformation of the roadway surrounding rock, but it has little impact on the changes in the stress and side abutment pressure of the roadway roof.

(4) **Comparison analysis of different roof cutting height scheme**

The D_1–D_5 scheme is compared and analyzed by using the roof cutting height as a variable. Figure 4.29a displays the side abutment pressure variation curves for the solid coal side and mine pillar in each scheme. Figure 4.29b displays the stress variation curve for the roadway roof. To comprehensively compare and analyze surrounding rock stress changes and deformation control laws of various schemes, various evaluation indexes are summarized, as shown in Fig. 4.30. In Fig. 4.30, the stress releasing rate of the roof is selected from the measurement points 3 and 7 of the roadway roof as typical representative points for shallow and deep roof evaluation.

By analyzing Figs. 4.29 and 4.30, it can be seen that:

1) **Analysis of the side abutment pressure**

When the cutting height is 4–12 m, the side abutment pressure rising rates of AFR by roof cutting are 56.2%, 46.4%, 36.1%, 32.9%, and 30.8%, respectively. The peak value of the side abutment pressure of the solid coal side continually drops and moves from about 3 m to approximately 6 m from the surface of the solid coal side when the cutting height is between 4 and 8 m. When the cutting height is 8–12 m, the solid coal side's side abutment pressure peak value is quite close to the surface, at a distance of around 6 m. According to the aforementioned findings, when cutting height grows, the peak side abutment pressure of the solid coal side steadily drops and progressively moves towards the deep. The peak side abutment pressure of the solid coal side remains relatively constant after the cutting height surpasses 8 m.

2) **Analysis of the roof pressure releasing effect**

With increasing from 4 to 8 m, stress releasing rate of the shallow measuring point 3 on the roof of AFR increases from 52.5% to 64.5%, and the stress releasing rate of the deep measuring point 7 increases from 11.2 to 22.6%. With increasing from 8 to 12 m, roof pressure releasing rate at the shallow measuring point 3 and the deep measuring point 7 does not increase significantly. This indicates that the larger the cutting height, the better the pressure releasing effect of the surrounding rock. The surrounding rock's ability to release pressure is not much enhanced after the roof cutting height reaches a particular point.

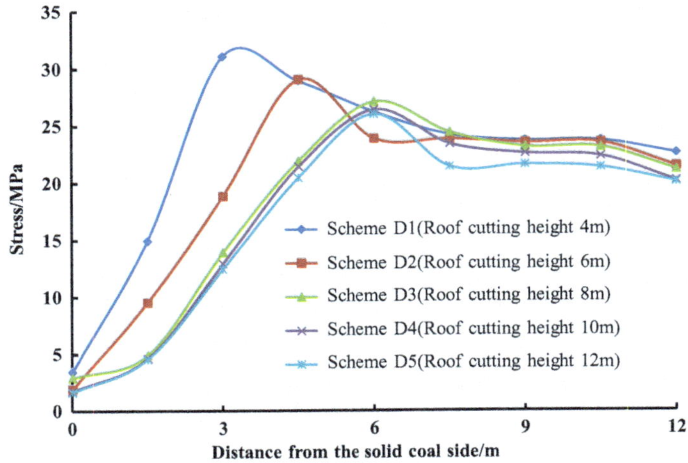

(a) Side abutment pressure with different roof cutting height

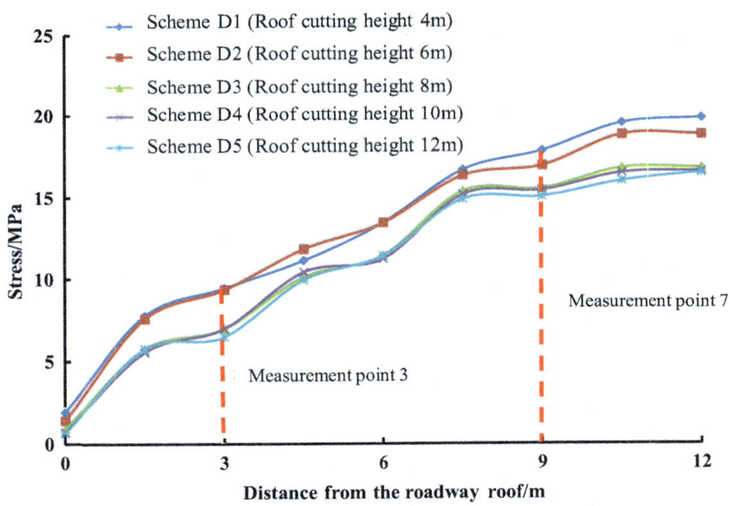

(b) Roof stress with different roof cutting height

Fig. 4.29 Variation of stress with different roof cutting height

3) **Analysis of the surrounding rock deformation control**

When the roof cutting height is 4–12 m, the deformation control rates of the surrounding rock of AFR are 0%, 28.4%, 43.9%, 50.2%, and 54.1%, respectively. The higher the cutting height, the higher the deformation control rate of the surrounding rock in the AFR. However, when the cutting height exceeds 8 m, the improvement

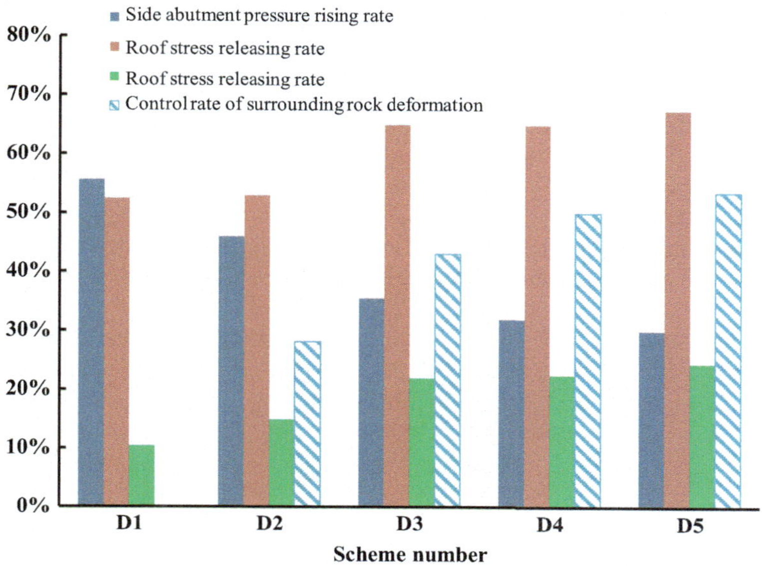

Fig. 4.30 Quantitative evaluation indexes of different cutting height

of the surrounding rock control effect is not significant. This indicates that when the cutting height is 4 m or 6 m, the collapsed gangue cannot fully fill the gob. After the cutting height exceeds 8 m, the collapsed gangue can fully fill the gob, and the roof of the roadway is effectively supported.

Based on the comprehensive analysis of scheme D, the following conclusions can be drawn. As the roof cutting height increases, the side abutment pressure, roof stress, and surrounding rock deformation of the solid coal side of the AFR continuously decrease. However, when the roof cutting height surpasses a specific limit, the enhancement of surrounding rock pressure releasing and control effect is no longer significant.

4.3 The AFR in Fault Structure Areas

The presence of fault geological structures exacerbates the difficulty of mining operations. To clarify the control effect of AFR mining method on the surrounding rock of fault structures, a comparative analysis is conducted between AFR and gob-side entry driving in fault structure.

4.3.1 Research Object

An accident involving a rock burst happened in 2020 at Xinjulong Mine. According to Fig. 4.31, the accident region is between 10 and 430 m from tail roadway to cut-off and between 0 and 66 m from the 3# crossheading to the tail roadway. The roadway is ruined across a wide region in this area due to excessive ground stress (horizontal stress can reach up to 47 MPa). The maximum convergence of the two sides is 2.8 m, while the maximum convergence deformation of the tail roadway's floor and roof is 2.4 m. The physical and mechanical characteristics of the main rock strata in the accident area are shown in Table 4.3.

FD8 fault has an average throw of 12 m and is mostly found in the 2305S working face. It has a 70° dip angle and a NW70° dip. The working face's tail roadway is a rectangular segment of 5100 × 4200 mm (width × height). The anchor cables that support the roadway roof are Φ22 × 6300 mm and Φ22 × 3800 mm in diameter.

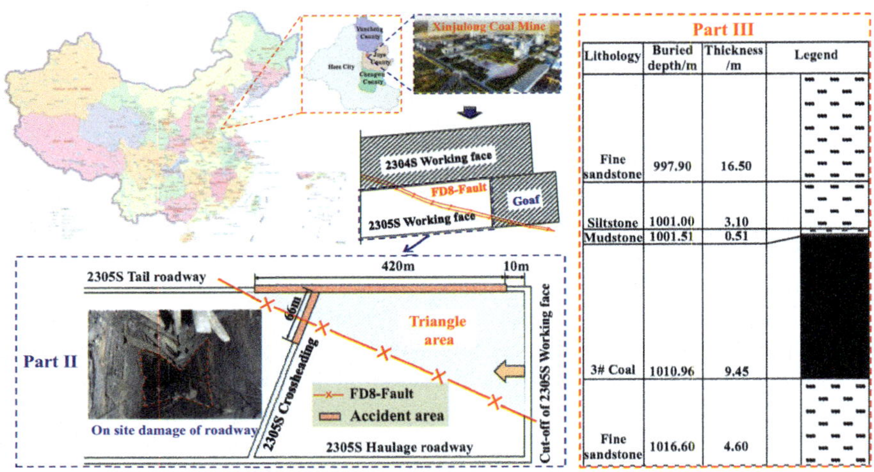

Fig. 4.31 The overall situation of the accident

Table 4.3 The physical and mechanical parameters of rock strata

Lithology	Elastic modulus/GPa	Poisson's ratio	Cohesion/ MPa	Internal friction angle/°	Tensile strength/MPa
Fine sandstone	54.43	0.20	5.76	34.5	3.99
Siltstone	32.10	0.21	5.09	34.5	2.14
Mudstone	19.68	0.23	4.42	35.0	2.09
Coal	21.60	0.30	2.88	28.5	0.74

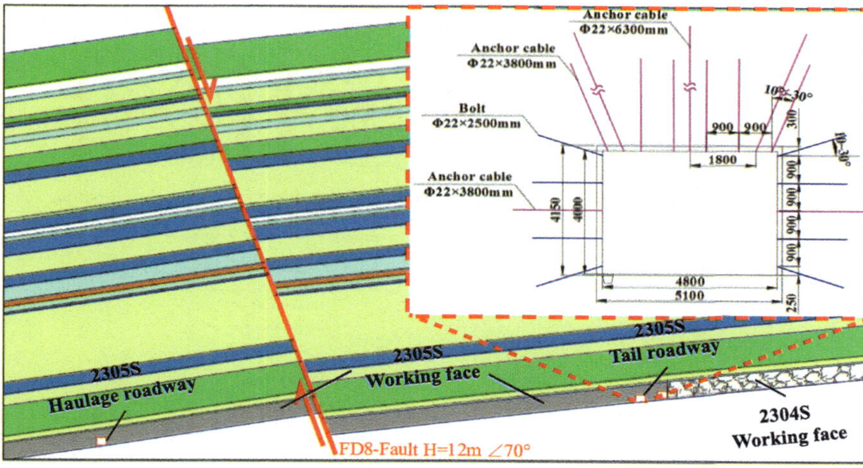

Fig. 4.32 Fault and roadway support parameters. Reprinted from Ref. Wang et al. (2022), Copyright © 2022, The Author(s), under exclusive licence to Springer Nature Switzerland AG

Both roadway sides use anchor cables and bolts as their support system. The bolt is $\Phi 22 \times 2500$ mm and cable is $\Phi 22 \times 3800$ mm. Figure 4.32 displays the particular fault characteristics and roadway support parameters.

4.3.2 Scheme Design

(1) **Overall scheme**

If the stress concentration of the deep roadway surrounding rock can be reduced by taking targeted pressure-releasing measures at stress concentration position, then risk of a disaster accident may be reduced or even cancelled. This is because the disaster accident of the deep roadway caused by the fault structure in the above typical engineering can be resolved. Therefore, a AFR technology for of deep mining roadway in fault structural areas is proposed. The technical process is shown in Fig. 4.33.

Numerical comparison test analysis on pressure releasing mechanism in deep roadway with fault structure is conducted to elucidate the influence mechanism of directional roof cutting on the pressure releasing effect.

In this section, the ground stress of the 2305S working face, the width of the mine pillar, the section size, the surrounding rock support strength of the tail roadway, and the thickness of the coal seam are considered as invariants. Two types of numerical comparison test techniques are developed to investigate variation stress of the roadway at different angles and heights of roof cutting.

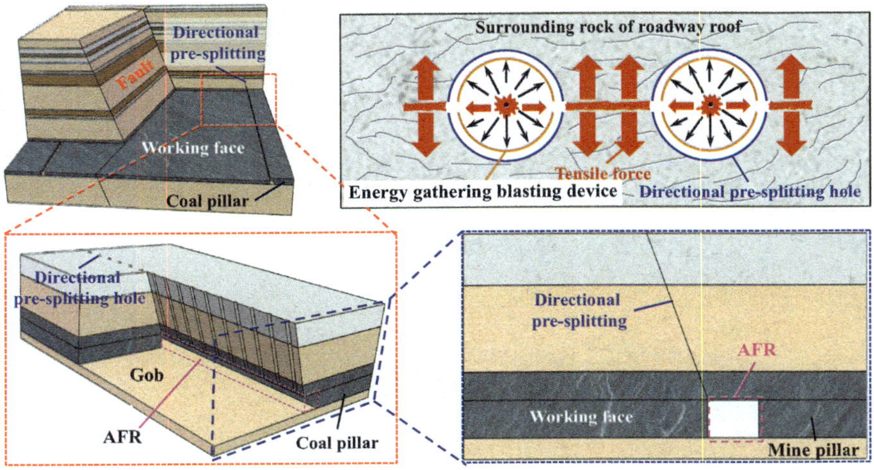

Fig. 4.33 The diagram of AFR mining method in fault structures. Reprinted from Ref. Wang et al. (2022), Copyright © 2022, The Author(s), under exclusive licence to Springer Nature Switzerland AG

1) Roof cutting height comparison scheme: The primary focus of this scheme's investigation is the roadway stress distribution and variation law with varied cutting height conditions. The roof cutting height is expressed as a variable, C_m, where m = 1–5. This translates to various heights of 15, 20, 25, 30, and 35 m, in that order.
2) Roof cutting angle comparison scheme: The primary focus of this scheme's investigation is the roadway stress distribution and variation law with varied cutting angle conditions. The roof cutting angle is expressed as a variable, D_n, where n = 1–5. This translates to various angles of 0°, 5°, 10°, 15°, and 20°, in that order.

(2) **Monitoring scheme**

The on-site geological conditions are taken into account when simulating a 440 × 192 × 240 m (width × height × thickness) range using FLAC^{3D} numerical software. The model, as shown in Fig. 4.34, adopts the Mohr–Coulomb constitutive model. The model's vertical displacement at its base is set to zero, and its horizontal displacements are limited to the normal direction. By placing compensatory tension on the model's top surface, the dead weight of underlying strata is mimicked.

A monitoring section is set at 180 m from the front surface of the model body. The stress monitoring points are arranged on the roadway roof and shoulder, which are denoted f and s (Ls represents left shoulder, Rs represents right shoulder), where f = 1–21, Ls = 1–11 and Rs = 1–11, respectively. The interval between the monitoring points is 2.0 m, as shown in Fig. 4.34. A row of stress monitoring points is set 20 m above the working face roof, with an interval of 9.0 m. The monitoring points of the working face roof are set to the left starting from the tail roadway roof and ending at the haulage roadway roof.

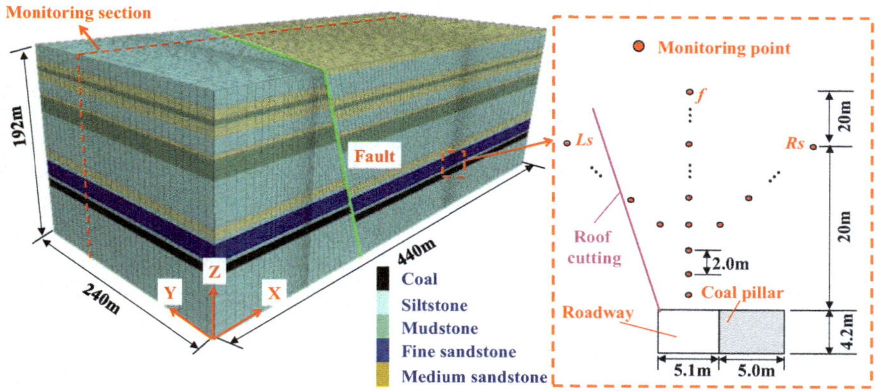

Fig. 4.34 The diagram of numerical model and monitoring scheme. Reprinted from Ref. Wang et al. (2022), Copyright © 2022, The Author(s), under exclusive licence to Springer Nature Switzerland AG

4.3.3 Result Analysis

The working face roof stress decreasing rate $\delta_{S_{ij}}$ and the roadway stress decreasing rate $\delta_{R_{ij-k}}$ are determined. This makes it possible to compare and analyze the stress fluctuation law and the pressure-releasing effects of various mining method more successfully.

1) The working face roof stress decreasing rate $\delta_{S_{ij}}$ represents percentage of working face's average roof stress that has decreases when compared to the working face without roof cutting. The following is the calculating equation:

$$\delta_{S_{ij}} = (\overline{S}_0 - \overline{S}_{ij})/\overline{S}_0 \times 100\% \tag{4.4}$$

where:

$\delta_{S_{ij}}$—The high stress influence area of scheme *ij* is the working face roof stress decreasing rate. Working face roof has a stronger pressure-releasing function the higher the $\delta_{S_{ij}}$ value. Among them, *i* represents the numbering of various schemes, where *i* = C–D, *j* represents the sub schemes of the general scheme, where *j* = 1–5. For instance, scheme C1's roof stress decreasing rate is denoted by $\delta_{S_{C1}}$, with a unit of %.

$\overline{S}_0$—Without roof cutting, the original roadway's working face roof's average stress of high stress influence area is measured in MPa.

$\overline{S}_0$—Roof average stress in scheme *ij*.

2) The percentage difference between average stress with roof cutting and one without is known as the roadway stress decreasing rate $\delta_{R_{ij-k}}$. The following is the calculating equation:

$$\delta_{R_{ij-k}} = (\overline{R}_{0-k} - \overline{R}_{ij-k})/\overline{R}_{0-k} \times 100\% \tag{4.5}$$

where

$\delta_{R_{ij-k}}$—The roadway's average stress decreasing rate in Scheme *ij*. There is less surrounding rock stress on the roadway with roof cutting in scheme *ij* compared to original roadway without roof cutting. The greater the pressure-releasing action of the surrounding rock roadway is, the higher the $\delta_{R_{ij-k}}$ value. *i* and *j* among them have the same significance as Eq. (4.4). Where $k = 1$ denotes the roadway encircling and $k = 2$ denotes the roadway shoulder, indicates the various types of roadway surrounding the rock. For instance, scheme C1's roadway roof stress decreasing rate is denoted by $\delta_{R_{C1-1}}$, with a unit of %.

$\overline{R}_{0-k}$—The average stress of the original roadway surrounding rock without roof cutting, MPa.

$\overline{R}_{ij-k}$—The average stress of the roadway surrounding rock in scheme *ij*, MPa.

(1) **Comparison analysis of working face stress distribution**

Stress distribution of working face in the fault parameter scheme (using the fault dip: NW 70°, dip angle: 70° as an example) and roof cutting parameter scheme (using the roof cutting height: 25 m, angle: 10° as an example) during the mining process is depicted in this Fig. 4.35. The vertical stress in the surrounding rocks is represented by MPa on the Z-axis, while the length of the working face mining is represented by Y and X, respectively.

By analyzing Fig. 4.35, it can be seen that:

The fault structure is clearly influencing the degree of stress concentration of the working face roof surrounding rock in the hanging wall area of the fault. The stress concentration range is lowered by the roadway roof cutting. In addition to clearly indicating the direction away from the roadway, the stress maximum location shifts from 18 to 36 m from the surface of the roadway. The findings demonstrate that directionally roof cutting of the roadway may significantly lessen the impact of stress concentration on the roadway and enhance the working face roof surrounding rock's stress environment.

Figure 4.36 displays the working face stress evaluating indexes and roof stress distribution under various roof cutting conditions.

By analyzing Fig. 4.36, it can be seen that:

1) Comparison of methods for releasing pressure: The two types of roof cutting schemes (scheme C and D) show a considerable decrease in working face advance stress when compared to no roof cutting scheme. Compared to the no roof cutting scheme, the maximum transfer distance can reach approximately 27 m. The stress peak location clearly moves to the direction that is farthest from the 2305S tail roadway.
2) Comparison of pressure releasing parameter schemes: As the roof cutting height rises, the advance stress peak point of the working face roof surrounding rock continuously moves. The decreasing rate of stress peak of the working face

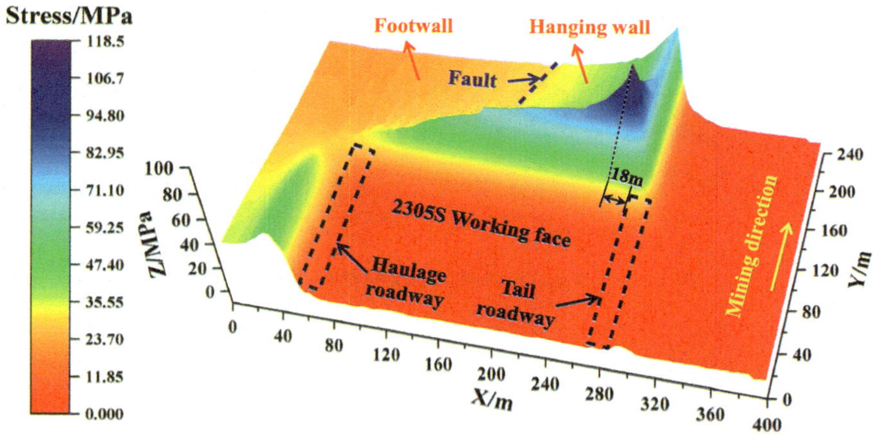

(a) Fault parameter scheme

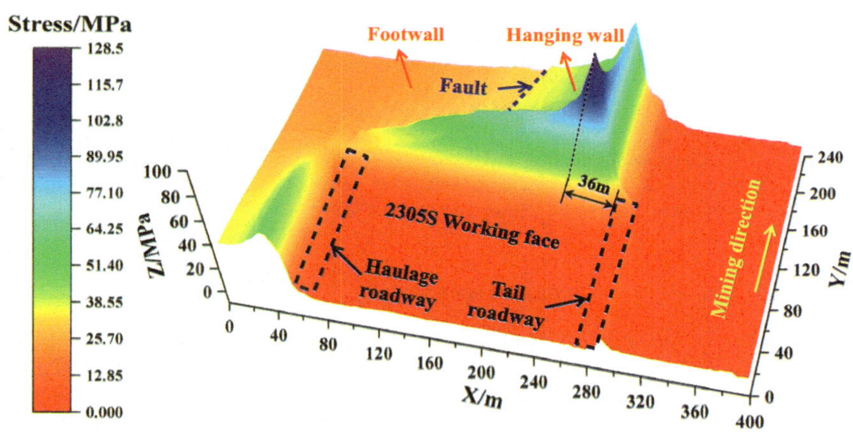

(b) Roof cutting parameter scheme

Fig. 4.35 Stress distribution in three dimensions of the working face roof. Reprinted from Ref. Wang et al. (2022), Copyright © 2022, The Author(s), under exclusive licence to Springer Nature Switzerland AG

roof is increasing. Advance stress peak location does not transfer clearly when the roof cutting angle increases. In scheme D4 (roof cutting angle: 15°), the pressure-releasing effect is most noticeable and the roof stress peak decreasing rate is the biggest.

(2) **Comparison analysis of roadway roof stress**

Figure 4.37 displays the pressure-releasing roadway's roof stress distribution and stress evaluation indexes under various roof cutting parameters.

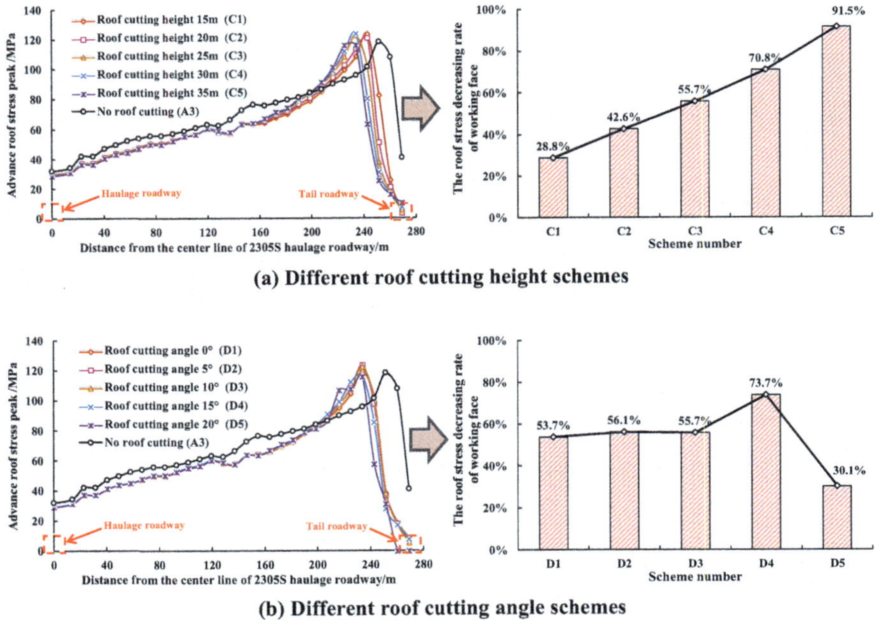

Fig. 4.36 The roof stress distribution and surrounding rock stress evaluation indexes of pressure releasing working face. Reprinted from Ref. Wang et al. (2022), Copyright © 2022, The Author(s), under exclusive licence to Springer Nature Switzerland AG

By analyzing Fig. 4.37, it can be seen that:

1) The area of the roadway roof affected by pressure release during roof cutting expands as the height of the roof cutting process increases. However, the area has not changed much despite the increased roof cutting angle. It demonstrates that roof cutting height has a bigger impact on the roof pressure-releasing effect than does roof cutting angle.
2) Of the roof cutting height schemes, scheme C4 (roof cutting height: 30 m) has the best pressure-releasing impact and the biggest roof stress decreasing rate (931.7%). Of the roof cutting angle schemes, scheme D2 (roof cutting angle: 5°) has biggest roof stress decreasing rate, at 852.5%. The aforementioned research indicates that the best pressure-releasing impact of the surrounding rock on the roadway roof may be achieved when the following roof cutting parameters are chosen: a height of 30 m and an angle of 5°.

(3) **Comparison analysis of roadway shoulder stress**

Figures 4.38 and 4.39 display the pressure-releasing roadway's shoulder stress distribution and stress evaluation indexes under various roof-cutting conditions.

By analyzing Figs. 4.38 and 4.39, it can be seen that:

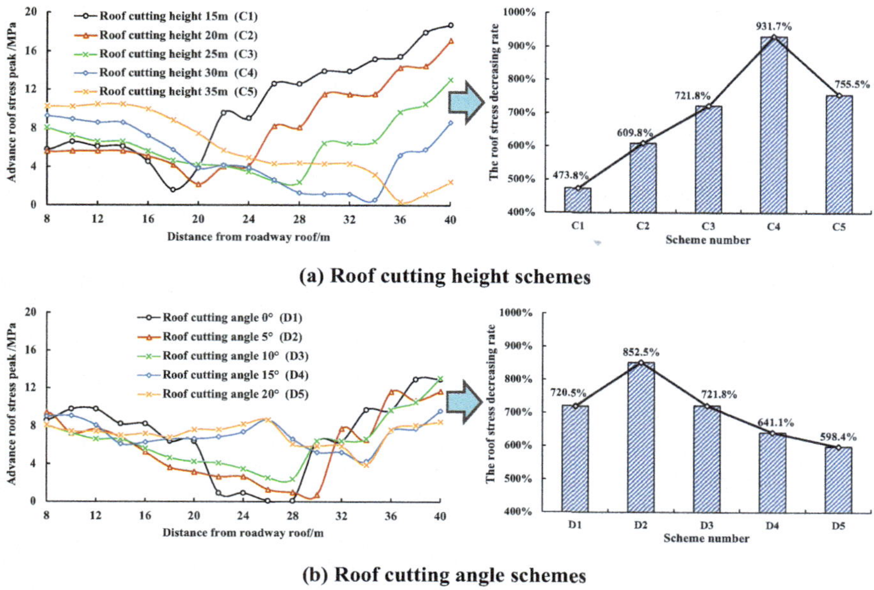

Fig. 4.37 The pressure-releasing roadway's surrounding rock stress evaluation indexes and roof stress distribution. Reprinted from Ref. Wang et al. (2022), Copyright © 2022, The Author(s), under exclusive licence to Springer Nature Switzerland AG

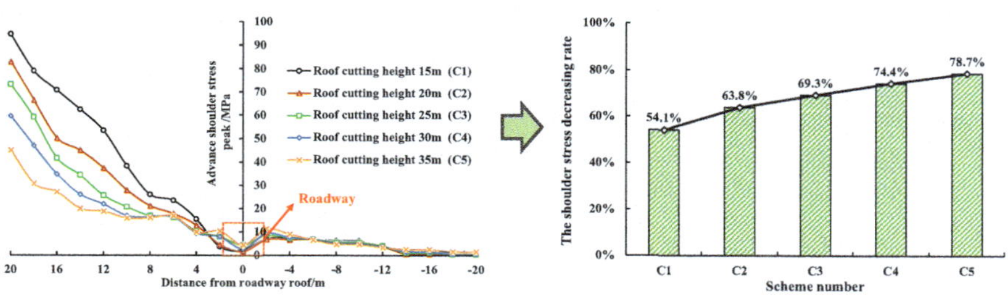

Fig. 4.38 The distribution of stress on the roof and the evaluation indexes of surrounding rock stress for various roof cutting height schemes. Reprinted from Ref. Wang et al. (2022), Copyright © 2022, The Author(s), under exclusive licence to Springer Nature Switzerland AG

1) Surrounding rock stress of the roadway left shoulder continuously decreases as the roof cutting height increases. The best and poorest schemes for releasing pressure have different stress releasing rates by 24.6%. However, the surrounding rock stress of the left shoulder of the roadway does not change significantly as the roof cutting angle increases. There is only a 9.3% difference in the stress releasing rate between the best and poorest pressure releasing effect schemes. It shows that roof cutting height, as opposed to roof cutting angle, has a greater influence on the roadway shoulder pressure release effect.

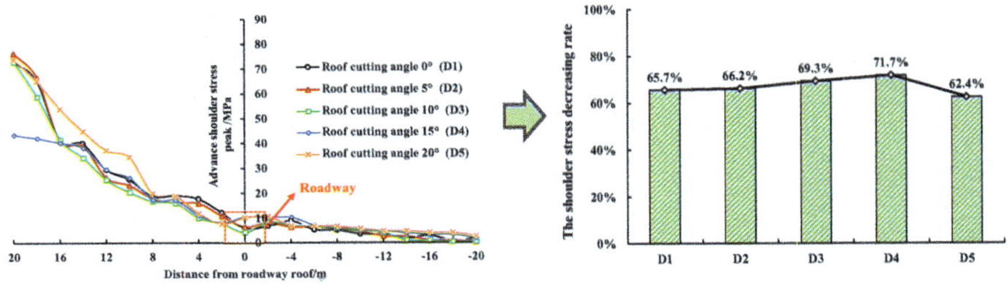

Fig. 4.39 The distribution of stress on the roof and the evaluation indexes of surrounding rock stress for various roof cutting angle schemes. Reprinted from Ref. Wang et al. (2022), Copyright © 2022, The Author(s), under exclusive licence to Springer Nature Switzerland AG

2) Of the roof cutting height schemes, scheme C5 (roof cutting height: 35 m) has the best pressure-releasing impact and the biggest roadway left shoulder stress releasing rate (78.7%). Scheme D4 (roof cutting angle: 15°) has the greatest roadway left shoulder stress releasing rate (71.7%) among the roof cutting angle schemes. Based on the aforementioned study, the optimal pressure-releasing impact of the surrounding rock on the roadway shoulder may be achieved by selecting the following roof cutting parameters: 35 m of height and 15°of angle.

The aforementioned study shows that roof directional cutting may successfully lessen impact of roadway stress concentration by enhancing working face and roadway stress environment. Closely spaced roof cutting heights of 25 and 35 m has an impact on the pressure release of the surrounding rock on the roadway roof. The appropriate range of roof cutting parameters may be established by taking into account the impact of two different types of roof cutting parameters on the surrounding rock's ability to release pressure and the construction efficiency factors. A appropriate range of roof cutting parameters, given the geological circumstances in this study, is a height of 25 to 30 m and an angle of 5° to 15°.

4.4 Chapter Summary

This study compares the traditional gob-side entry driving with mine pillar method and AFR mining method in numerical simulation. It addresses the control problem of AFR under shallow buried depth, deep buried depth, and fault structural areas. The progression law of roadway stress and the distribution characteristics of stress in the mining region are analyzed, and the foundation for roof cutting and support parameters are clarified.

References

Wang Q, Wang Y, He MC, Jiang B, Li SC, Jiang ZH, Wang YJ, Xu S (2021) Experimental research and application of automatically formed roadway without advance tunneling. Tunn Undergr Space Technol 114:103999

Wang Q, Wang Y, He MC, Li SC, Jiang ZH, Jiang B, Xu S, Wei·HY (2022) Experimental study on the mechanism of pressure releasing control in deep coal mine roadways located in faulted zone. Geomech Geophys Geo-Energy Geo-Resources 8(2):1–24

Chapter 5
Geomechanical Model Test of AFR Mining Method

Geomechanical model testing is an effective means of simulating the overall research of engineering sites, but existing model testing systems do not have specialized equipment to simulate the core process of AFR mining methods. Develop a three-dimensional geomechanical model test system that can truly simulate the entire process of mining, excavation, and retention integration using the AFR mining method without pillars. A comparative test study is conducted on the model of AFR mining method and gob-side entry driving method.

5.1 Model Test System of AFR Mining Method

5.1.1 System Development Concept

The new model test system has to fulfill the following requirements in order to guarantee a genuine simulation of the AFR mining method without mine pillars.

(1) Key technologies for AFR mining method should be created in accordance with the fundamental procedures of AFR mining method without mine pillars, which include mining, roof cutting, and roadway formation. In order to accomplish automated form roadway and ensure consistency between the model test and on-site AFR technology, simultaneous roof cutting and gangue support are performed at each footage of the working face after the mining is completed.
(2) The test system has to have enough loading capacity and simulation area to replicate the whole mining operation as well as the automatic form of a stable roadway. Under circumstances like initial pressure and periodic pressure, it may carry out study on the influence mechanism of mine pressure.

Q. Wang et al., *Theory and Method of Automatically Formed Roadway in Underground Mining*, https://doi.org/10.1007/978-981-97-8082-2_5

(3) Multiple cross-sections and measurement locations can have high precision displacement and stress monitoring devices deployed inside the model. Throughout the whole testing procedure, real-time monitoring of mining pressure monitoring data, including overlaying rock movement, stope stress, and roadway convergence and so on, is conducted.

5.1.2 Test System Introduction

Roof cutting, roadway form, mining, and other basic technologies make up the majority of the AFR mining technology that is essential to the model test. The main components of coal mining are support simulation, automatic excavation system, and roof directional cutting. The specific composition of the model testing system is shown in Fig. 5.1.

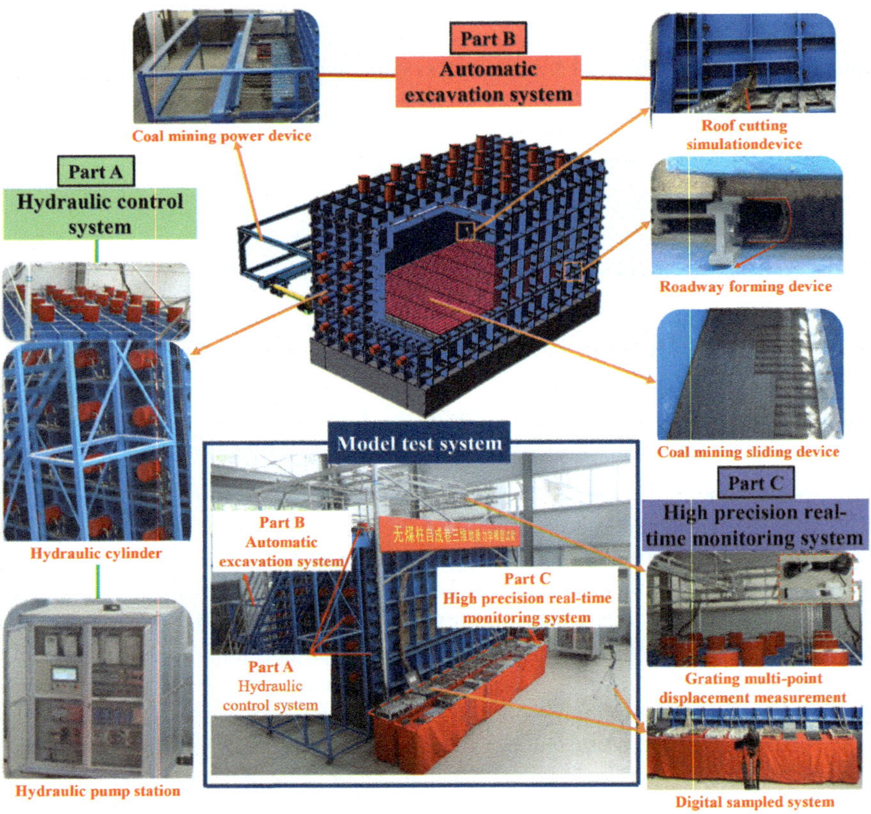

Fig. 5.1 Three dimensional geomechanical model test system for AFR

A set of geomechanical model test systems for AFR mining method is developed. These systems are primarily used to simulate AFR under various situations without the usage of mine pillars. The hydraulic control system, automatic excavation system, and high precision real-time monitoring system comprise the test system. The model test system is shown in Fig. 5.1.

1) **Hydraulic control system**

Part A of Fig. 5.1 shows the hydraulic control system, which consists of the hydraulic cylinder and hydraulic pump station. The system consists of four reaction frames, reinforced ribs, connecting members, front and rear reaction walls, and windows composed of high-strength tempered glass. Model bodies of different sizes up to 5.5 m × 2.4 m × 3.0 m (width × height × thickness) can be tested and researched. The reaction device can be as large as 6.4 m × 4.4 m × 3.4 m (width × height × thickness).

To reduce friction and the boundary effect, a PTFE plate is positioned between the model body and the front & rear reaction walls. The front reaction wall's center also houses the tempered glass window. During the mining operation, the movement of the surrounding rock and the deformation of the roadway may be seen via the window.

2) **Automatic excavation system**

As illustrated in Part B of Fig. 5.1, the automatic excavation system during mining consists of the mining sliding device, roof cutting simulation device, roadway forming device, and mining power device.

i) The top separator, bottom separator, connecting parts, and bearing sliding unit make up the mining sliding device, which is intended to simulate mining. The upper separator is made up of many unit plates and has a divided pattern. It collapses as a result of the bearing sliding unit losing support, simulating the collapse of the underlying strata and mining.
ii) With its various roof cutting plates, the roof cutting simulation device is intended to simulate both roof cutting and mining. The entire apparatus is made up of the roof cutting plates' head and tail. To guarantee that roof cutting is implemented effectively, the device's surface is coated with antifriction materials that can lower the friction resistance between the device and the model rock strata.
iii) The roof cutting simulation device and the mining sliding device are precisely controlled by the mining power device. It consists of an infrared sensor, a longitudinal cylinder, a transverse pull rod, a longitudinal connection, and a response support structure. On the model reaction support structure, the longitudinal cylinders are fixed. With the aid of the longitudinal connection, the cylinder pulls the transverse pull rod, causing the mining sliding device and the roof cutting simulation device to move outward simultaneously. To guarantee a realistic replica of the mining cycle film, the movement distance is controlled by the infrared sensor.

3) **High precision real-time monitoring system**

The high-precision real-time monitoring system consists of three different monitoring systems, which including digital photographic displacement monitoring, fiber optic strain monitoring, grating multi-point displacement monitoring systems and so on. These systems monitor the entire process of roadway convergence, mining area stress, and rock strata movement in real-time.

i) The digital photographic displacement monitoring system consists of picture analysis software and digital camera hardware. Images are captured using digital photography equipment, and image analysis software is then used to follow the measurement point's position and determine its displacement.
ii) The optical fiber strain monitoring system is made up of an optical fiber coupler, grating demodulation device, and acquisition software. The optical fiber strain sensor collects the monitoring signal, which is then connected via optical cable to the optical fiber coupler. The optical fiber grating demodulator is attached to the input end of the optical fiber coupler. The monitoring program converts the signal in order to gather data on the surrounding rock's stress and strain.
iii) A measuring point, measuring wire, separator, protective tube, grating ruler, steering wheel, balancing block, acquisition instrument, and acquisition software make up the grating multi-point displacement monitoring system. The steering wheel, the measuring wire, and the measuring point all link the measuring point to the balancing block. A separator is inserted within the protective tube to separate the measuring wire, which is protected by the protective tube. Through the measuring wire, the deformation is conveyed to the grating ruler and transformed into photoelectric signals (Fig. 5.2).

(3) **Test system function and performance index**

1) Test system function
    i) The test system can realize the coal mining, automatic roof cutting and AFR, the synchronous support of the support, and the real simulation of the "mining, roadway driving and roadway retaining" process flow of the AFR mining method.
    ii) The test system can carry out tests on the model body with the maximum size of 5.5 m × 2.4 m × 3.0 m, and monitor and analyze the overburden movement and mine pressure behavior law in the whole process of mining.
    iii) The test system has a combined reaction device, which can adjust the number of reaction frames according to different test requirements, and carry out geomechanical model tests of different model sizes.

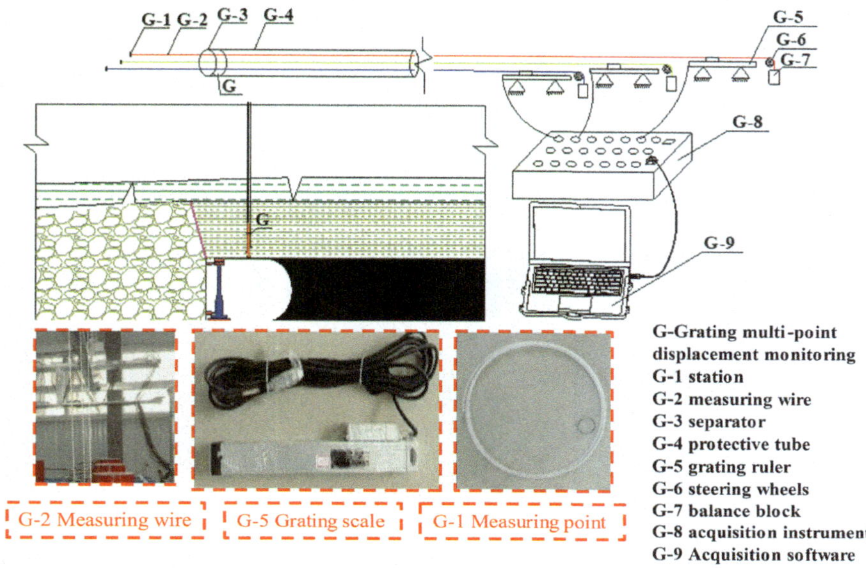

Fig. 5.2 Grating multi-point displacement monitoring system

iv) The test system can simulate mining with the AFR mining method without mine pillar, and can also simulate the traditional mining method.

2) Test system performance index

- The main size of the system: 6.4 m × 4.4 m × 3.4 m (width × height × thickness);
- Model body size: 5.5 m × 2.4 m × 3.0 m (width × height × thickness);
- Hydraulic cylinder loading: load concentration 5.0 MPa, accuracy ± 0.01%F.S;
- Mining: speed 20 ~ 50 mm/min, accuracy 1 mm/min;
- Grating multi-point displacement monitoring: measuring range ± 250 mm, accuracy 0.1 μm;
- Fiber strain monitoring: measuring range ± 5000 με, accuracy 1 με.

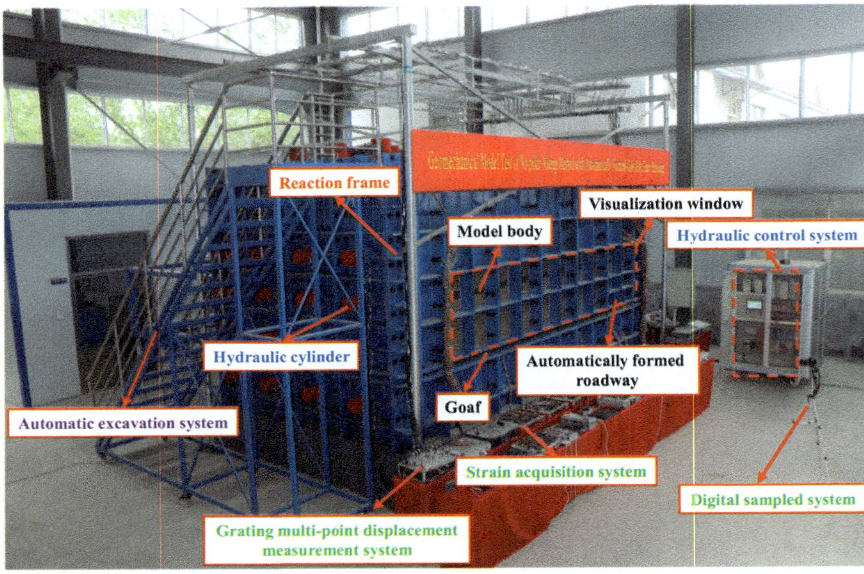

Fig. 5.3 The model test of the AFR with shallow buried depth. Reprinted from Ref. Wang et al. (2020), Copyright © 2020 Elsevier Ltd.

5.2 The AFR with Shallow Buried Depth

5.2.1 Research Object

A three-dimensional geomechanical model test system is utilized to conduct comparative model test study in order to confirm the pressure relief impact of the AFR mining method and gob-side entry driving method with shallow buried depth. The geological conditions and physical and mechanical parameters of rock mass are in line with Sect. 4.1.1, and it is carried out on the Ningtiaota Mine's working face with shallow buried depth. The overall situation of the specific model test is shown in Fig. 5.3.

5.2.2 Scheme Design

(1) **Model parameters**

The bulk density similarity scale C_γ is 1.4 and the geometric similarity scale C_l is 60, based on the field geological conditions and model test system.

The geological range of 330 m × 144 m × 180 m is chosen for simulation based on the actual conditions on-site, and the converted model size is 5.5 m × 2.4 m × 3.0 m. Figure 5.4 displays the general layout of the geomechanics model test.

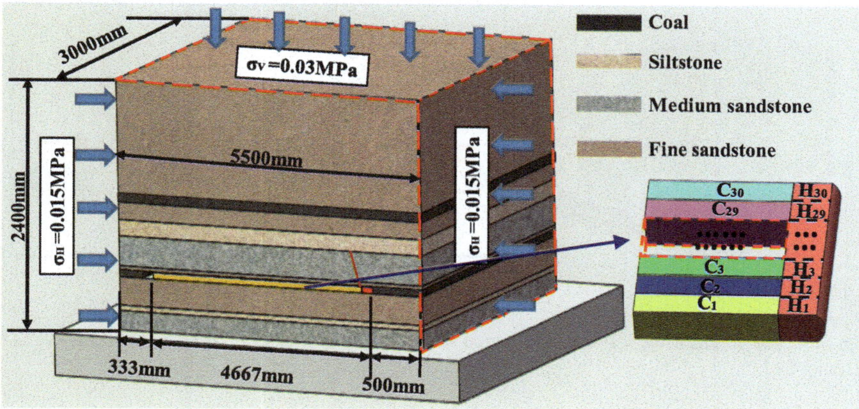

Fig. 5.4 The model design scheme. Reprinted from Ref. Wang et al. (2020), Copyright © 2020 Elsevier Ltd.

(2) **Loading and excavation scheme**

Three stages comprise the simulation scheme for the whole process: the loading stage, the AFR formation stage, and the AFR reuse stage.

1) The loading stage

The load imposed on the model surface is calculated to be 0.03 MPa in a vertical direction and 0.015 MPa in a horizontal direction based on the findings of the ground stress test. The hydraulic control system loads the model's upper surface as well as its two sides. Mining the working face and digging the roadway will occur once the load stabilizes and reaches the intended ground stress value.

2) The AFR formation stage

Firstly, using the automatic excavation system, left roadway is excavated for 100 mm by excavation footage, followed by a 20 min stabilization period. Secondly, the roof cutting plate of the right AFR is pulled out for 100 mm along the gob-side to simulate the roof-cutting of the AFR. Thirdly, the working face is excavated for 100 mm of excavation footage each time after holding for a 20 min stabilization period. Finally, repeat second and third step after each excavation and stabilization circle.

3) The AFR reuse stage

According to the reference, depending on the stage of mining, the stress concentration coefficient might increase to two or three times the initial rock stress. The overload stress concentration coefficient increases by 0.25 every 20 min during the test, starting at 1.25 and continuing until it reaches 3.0.

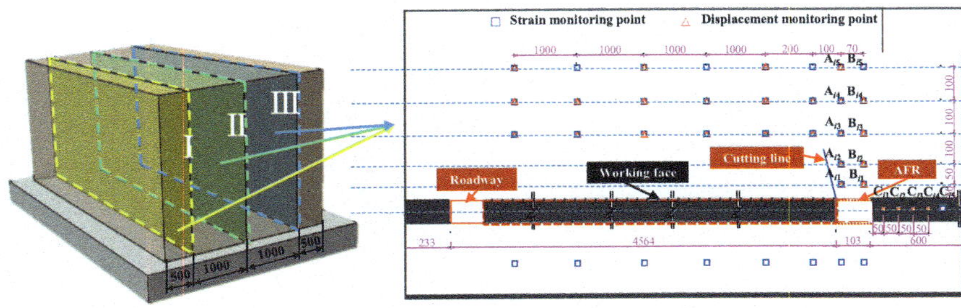

(a) Monitoring section (b)Monitoring position

Fig. 5.5 The monitoring scheme. Reprinted from Ref. Wang et al. (2020), Copyright © 2020 Elsevier Ltd.

(3) Monitoring scheme

Three monitoring sections are set up inside the model along the strike of the simulated working face in order to monitor the stress and deformation law of the surrounding rocks at the AFR. Figure 5.5a depicts the monitoring sections I, II, and III. The distances between the surfaces of the three monitoring portions and the model body are, respectively, 500 mm, 1500 mm, and 2500 mm. These parts have components for measuring stress and displacement placed on them. The two kinds of points are spaced 100 mm apart. The arrangement of the monitoring components is as follows: With i standing for the monitoring section number, $i =$ I–III, and j for the monitoring number, $j = 1$–5. A_{i-j} is located near the gob-side of the AFR roof, B_{i-j} is located near the solid coal side of the AFR roof, and C_{i-j} is located at the solid coal side. A detailed layout is presented in Fig. 5.5b. In order to picture the model's surface, a digital sampled system is also set up in front of the model test system.

5.2.3 Test Process

The specific excavation process of the model test is shown in Fig. 5.6.

5.2.4 Result Analysis

(1) Analysis of the AFR formation stage

1) The surrounding rocks stress evolution law
   a) Roadway roof surrounding rock stress

As the working face is mined, the stress law of the surrounding rocks on the roof of gob-side in monitoring section I (scheme $A_{I\text{-}1}$–$A_{I\text{-}5}$) is monitored. In Fig. 5.7, the stress variation curves are displayed.

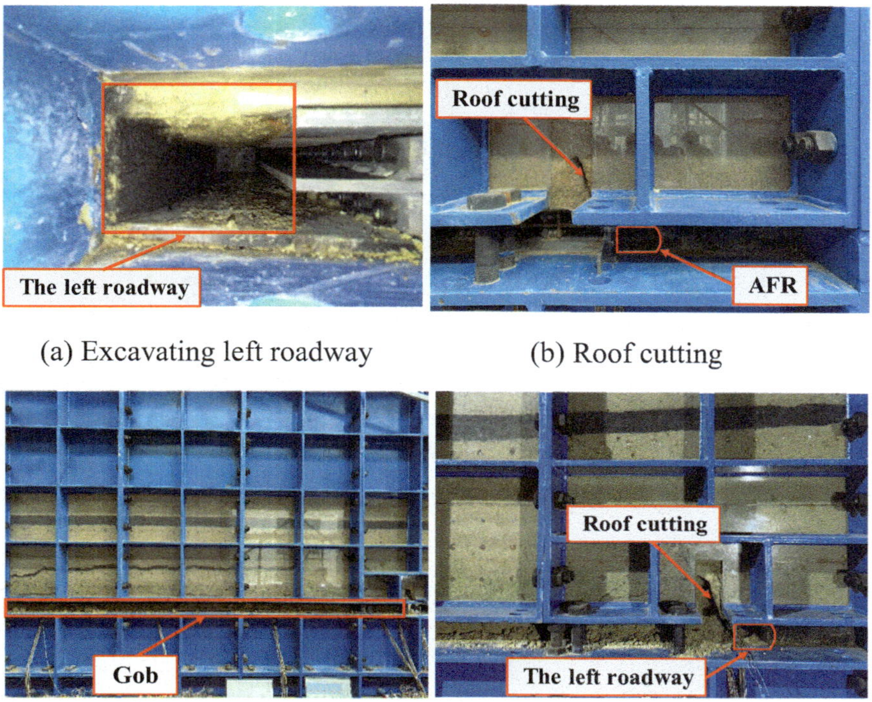

(a) Excavating left roadway (b) Roof cutting

(c) Excavating working face (d) Forming the AFR

Fig. 5.6 Geomechanical model test process. Reprinted from Ref. Wang et al. (2020), Copyright © 2020 Elsevier Ltd.

The research demonstrates that when the working face is pushed toward the monitoring section, the advanced roof stress steadily rises. The peak of the roof stress is reached when the working face lags by approximately 400 mm. The surrounding rock stress on the roadway roof is continuously relieved and redistributed as the working face advances. The stress is essentially steady and the stress recovery region finally forms when the working face lags by approximately 1000 mm.

b) Roadway solid coal side surrounding rock stress

In monitoring section I (scheme $C_{I\text{-}1}$ to $C_{I\text{-}5}$), side abutment pressure varies in tandem with the mining of the working face. In Fig. 5.8, the variation curves are displayed.

According to the analysis, the solid coal side stress steadily increases when the working face is mined. Stress values at each monitoring site increase to 0.079, 0.072, 0.067, 0.055, and 0.049 MPa as the working face is mined to the monitoring section I. The peak of the side abutment pressure of the surrounding rocks progressively shifts to the deep as a result of the working face mining's deformation of the roof rock mass in the gob and the action of roof cutting. After the working face is mined by

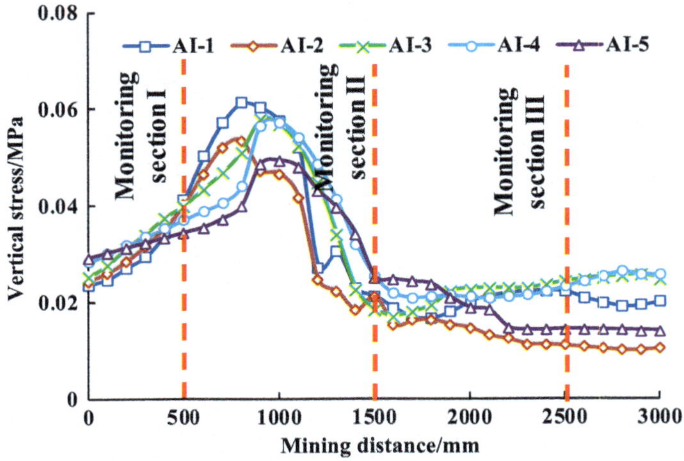

Fig. 5.7 The roadway gob-side roof stress. Reprinted from Ref. Wang et al. (2020), Copyright © 2020 Elsevier Ltd.

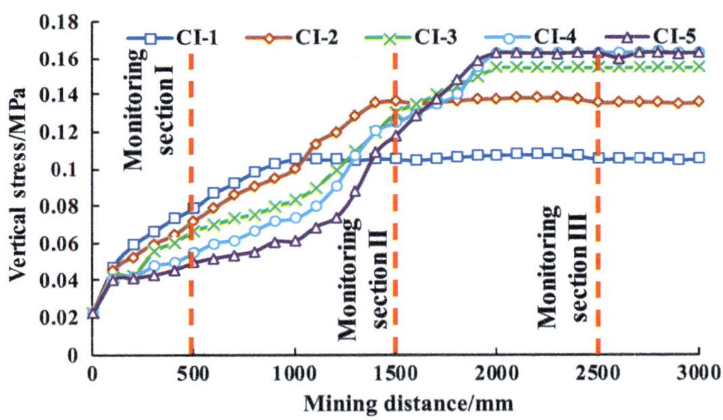

Fig. 5.8 The roadway solid coal side stress. Reprinted from Ref. Wang et al. (2020), Copyright © 2020 Elsevier Ltd.

1500 mm, the shift speed of the side abutment pressure peak slows down, and at this point, the peak pressure is 0.163 MPa. The collapsed gangue gets compressed and stabilized as the working face lags by 2000 mm, while the stress of the surrounding rocks remains mostly unaltered at the solid coal side.

2) The surrounding rocks internal displacement

The displacement of surrounding rocks on the solid coal side (scheme B_{I-1} to B_{I-5}) and the roof of the gob-side (scheme A_{I-1} to A_{I-5}) varies in tandem with the mining of the working face. The curves of variation are displayed in Fig. 5.9.

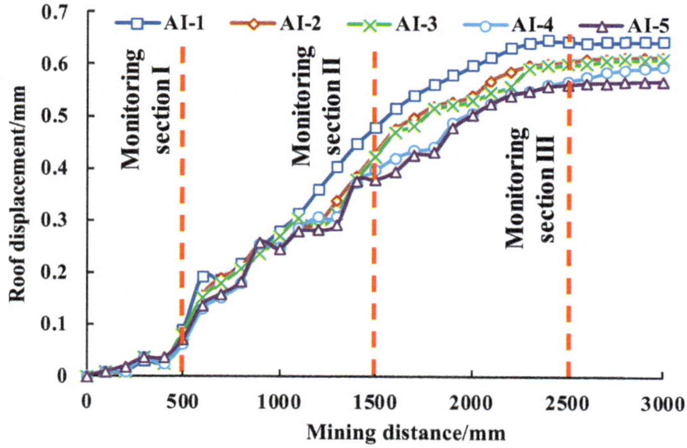

(a) Roof of gob-side

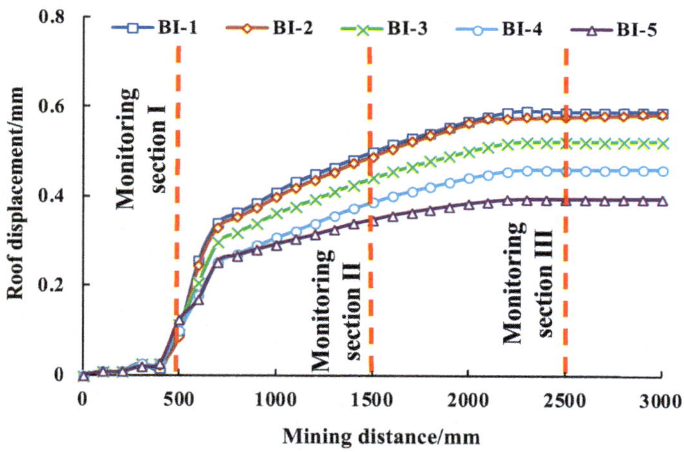

(b) Roof of solid coal side

Fig. 5.9 The roadway roof displacement of AFR. Reprinted from Ref. Wang et al. (2020), Copyright © 2020 Elsevier Ltd.

According to the investigation, the gob-side roof displacement is more noticeable than the solid coal side. It suggests that in the dynamic pressure area of the lagged working face, the cantilever end of the short cantilever beam is more susceptible to deformation. Furthermore, a greater displacement at the roof cutting side is caused by the friction and disruption of the falling gangue on the roof cutting surface. Compared to the deep measurement point, the shallow measuring point has a greater displacement. Falling gangue eventually fills the gob along with the mining of the

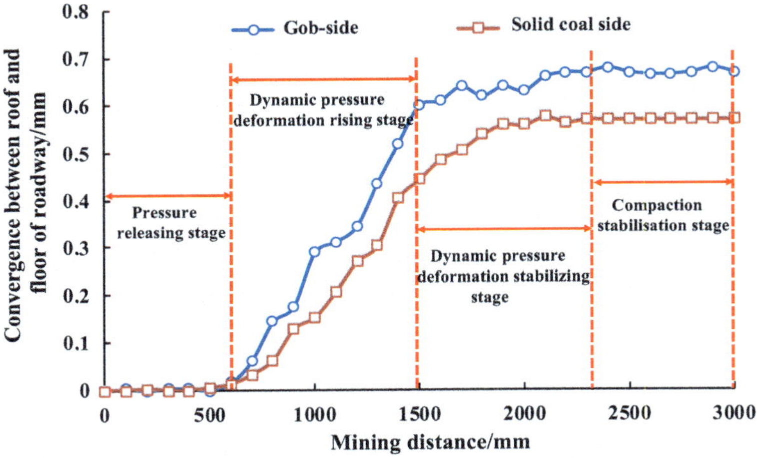

Fig. 5.10 The roof and floor convergence deformation of AFR. Reprinted from Ref. Wang et al. (2020), Copyright © 2020 Elsevier Ltd.

working face, stabilizing the roof's sinking over time. The roof surrounding rock has good stability and meet safety criteria. The maximum subsidence of the roof is 0.65 mm.

3) The roadway surrounding rock deformation

Figure 5.10 displays the variation curve for the AFR's roof and floor of convergence deformation.

The roadway roof and floor convergence deformation proceeds through the following four distinct times, as shown in Fig. 5.10.

a) The pressure releasing stage: The working face is lagging between 0 and 600 mm, the AFR roof and floor convergence deformation is minimal. This is primarily because the roof's collapse to the gob has little effect on the AFR roof.
b) The dynamic pressure deformation rising stage: The AFR's roof and floor convergence deformation grows quickly as working face lags from 600 to 1500 mm. Roof rotates and sinks due to the mining effect, the gob's roof rock mass is in a relatively strong motion condition, and the subsidence speed is quicker, which causes the AFR's surrounding rocks to clearly deform.
c) The dynamic pressure deformation stabilizing stage: The AFR's roof and floor convergence deformation speed slows down as the working face lags from 1500 to 2300 mm. The dynamic mining pressure causing the gangue to get compacted is still affecting the roof at this point, which means it has not fully stabilized. The surrounding rocks on the AFR tend to deform steadily as the gangue progressively compacts, increasing its support capacity on the strata above.

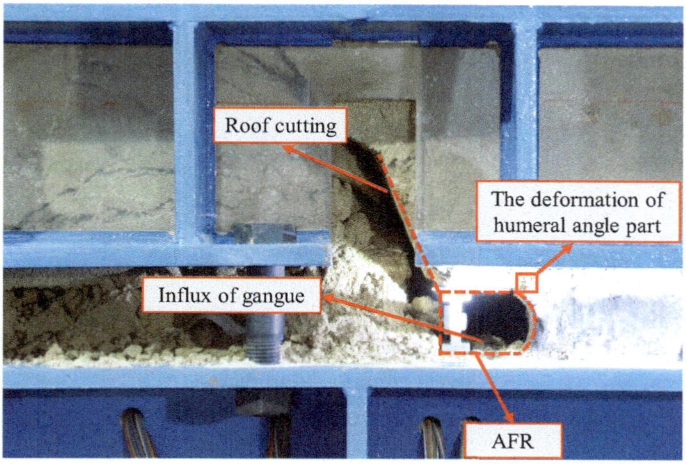

Fig. 5.11 The AFR deformation state in the overload test. Reprinted from Ref. Wang et al. (2020), Copyright © 2020 Elsevier Ltd.

d) The compaction stabilization stage: The rock mass around the working face essentially stabilizes after the working face lags by 2300 mm. The roof and floor convergence deformation exhibits little deformation, and the rate of deformation is also considerably slowed down. At this point, the surrounding rocks of the AFR are at the compaction stabilization stage, and the falling gangue has essentially been compressed in the gob.

(2) **Analysis of the AFR reuse stage**

An overload test is conducted to more clearly identify the important damaged areas of the AFR and the surrounding rocks' deformation law. It replicates the impact of mining during the AFR's reuse stage, and Fig. 5.11 displays the test result. Figure 5.12 displays the variation curves of the surrounding rock deformation during the overload test.

The analysis is shown as follows. As the overload stress increases, the humeral angle portion of the roadway side starts to crack, but the total deformation of the AFR surrounding rock is relatively modest. The humeral angle portion of the roadway side and the gob-side are the main damaged areas, as evidenced by the concurrent influx of gangue from the gob-side.

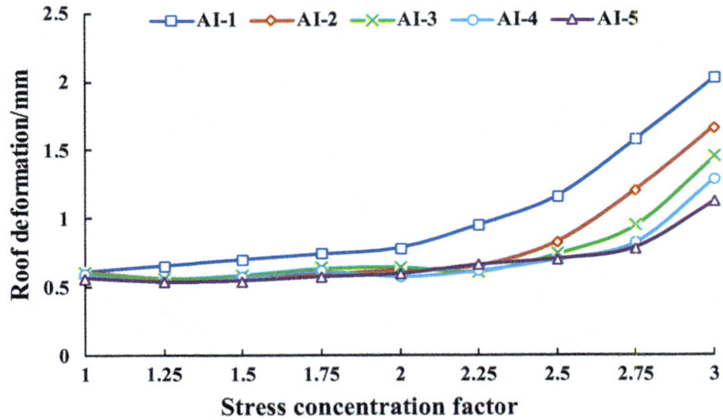

(a) Roof of gob-side

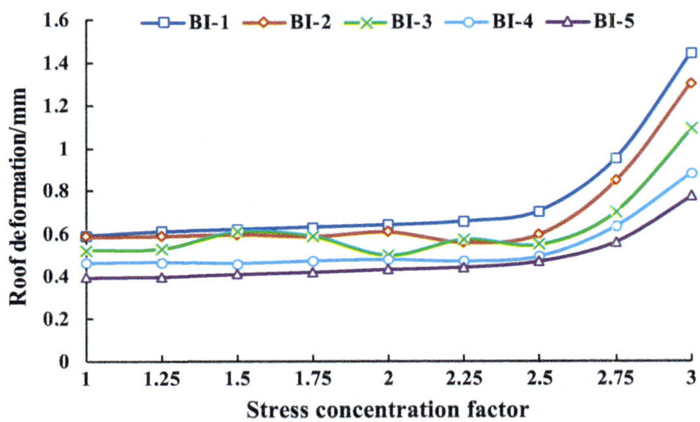

(b) Roof of solid coal side

Fig. 5.12 Roof surrounding rock deformation in the overload test. Reprinted from Ref. Wang et al. (2020), Copyright © 2020 Elsevier Ltd.

5.3 The AFR with Deep Buried Depth

5.3.1 Research Object

A three-dimensional geomechanical model test system is utilized to conduct comparative model test study so as to confirm the pressure relief impact of the AFR mining method and gob-side entry driving method with deep buried depth. The geological conditions and physical and mechanical parameters of rock mass are in line with

Fig. 5.13 The model test of the AFR with deep buried depth. Reprinted from Ref. Wang et al. (2021), Copyright © 2021 Published by Elsevier B.V. on behalf of China University of Mining & Technology

Sect. 4.2.1, and it is carried out on the Suncun Mine's working face with deep buried depth. The overall situation of the specific model test is shown in the Fig. 5.13.

5.3.2 Scheme Design

(1) **Model parameters**

Utilizing a test system, the model body measuring 2400 mm × 2400 mm × 500 mm (width × height × thickness) is tested. Figure 5.14 displays the model body size of model test.

The geometric similarity scale of the model is found to be 30. It is found that the stress similarity scale is 45 and the bulk density similarity scale is 1.5.

In the model, the size of AFR is 127 mm × 83 mm. The AFR's roof cutting height is 267 mm in the model and 8 m in the engineering field. The roof cutting angle in both the field and the model is 20°, and mica sheets are pre-buried at the AFR roof cutting positions in the model. In the model, the mine pillar width is 167 mm, however, it is 5 m in the field. The driving size strategy for gob-side entry driving is the same as for the AFR.

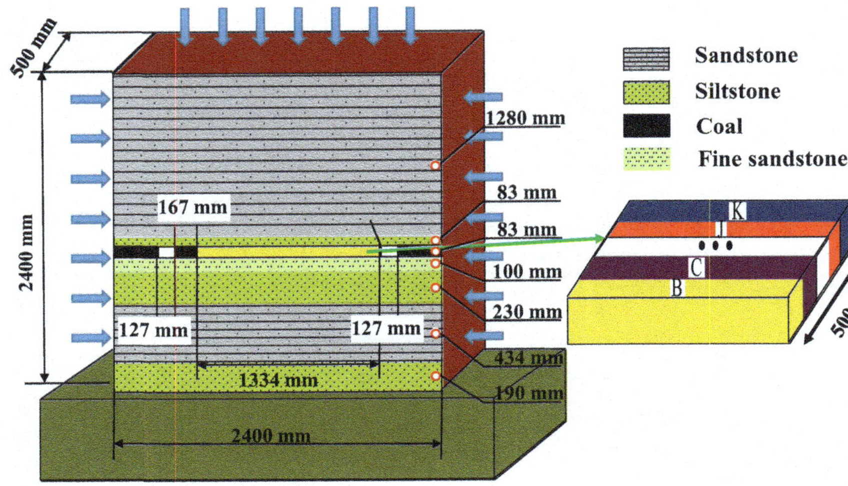

Fig. 5.14 Model body size of model test. Reprinted from Ref. Wang et al. (2021), Copyright © 2021 Published by Elsevier B.V. on behalf of China University of Mining & Technology

(2) **Loading and excavating scheme**

The model loading device is utilized to load different stress levels on the upper surface and left & right sides of the model. Through on-site measurement of in-situ stress and similar scales, determing the load concentration that needs to be applied to the surface of the model as 0.52 MPa in the horizontal direction and 0.43 MPa in the vertical direction.

The simulated roadway excavation sequence is as follows: excavation in the order A → B–K → L. That is, excavating AFR by roof cutting; middle working face; and left side gob-side entry driving. Figure 5.15 shows the specific excavation scheme.

1) Ten steps, from A_1 to A_{10}, are dug in the right side AFR by roof cutting. Each step is 50 mm.
2) A circuitous excavation is conducted in the direction of the working face while the intermediate mining face is dug in ten groups of B_i to K_i ($i = 1$–8). Each step is 50 mm.
3) The excavation mode used for the AFR by roof cutting is the same as the gob-side entry driving on the left side, which is split into ten groups from L_1 to L_{10}. Each step is 50 mm.

Overload tests are performed on the finished excavation model so as to mimic the effects of the next working face mining on the previously indicated gob-side entry driving and AFR. According to research, mining can cause adjacent strata to collapse, and the effect on rock stress next to a mine wall in a roadway can be up to three times greater than the initial rock stress. To simulate the impact of secondary mining, 2.5 times the overload coefficient is chosen in this work.

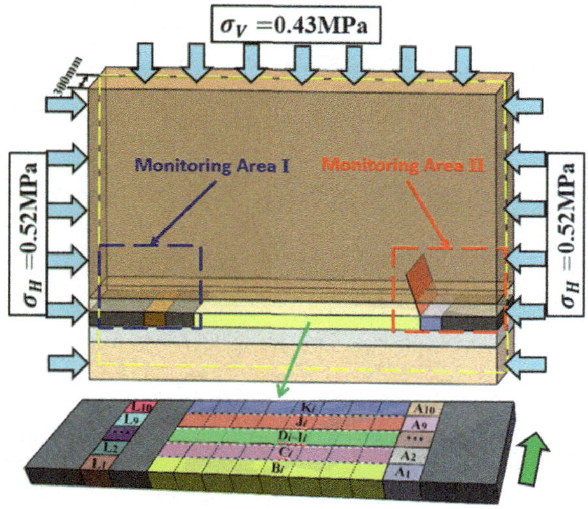

Fig. 5.15 The model test scheme

(3) **Monitoring scheme**

The purpose of monitoring sections I and II is to clarify the surrounding rock stress and deformation laws as well as mining. They are positioned in the model body at 100 mm and 300 mm from the model front surface, respectively. Nine monitoring lines are arranged in two monitoring sections to facilitate the deployment of the monitoring sites. The names of the monitoring points for the two monitoring sections are A_{i-j} and B_{i-j} (where i stands for monitoring line and j for monitoring point). The exact monitoring and positioning of the sections are shown in Fig. 5.16, where monitoring section I is used as an example. Moreover, a digitally sampled device is placed directly in front of the model body's front surface so as to conduct real-time image monitoring of the model surface.

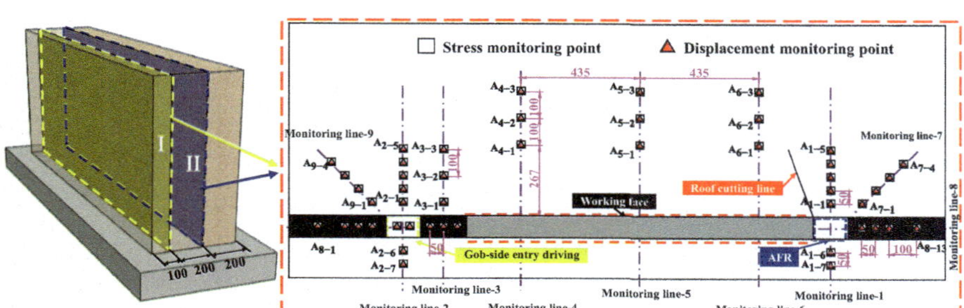

Fig. 5.16 The monitoring scheme in the model test. Reprinted from Ref. Wang et al. (2021), Copyright © 2021 Published by Elsevier B.V. on behalf of China University of Mining & Technology

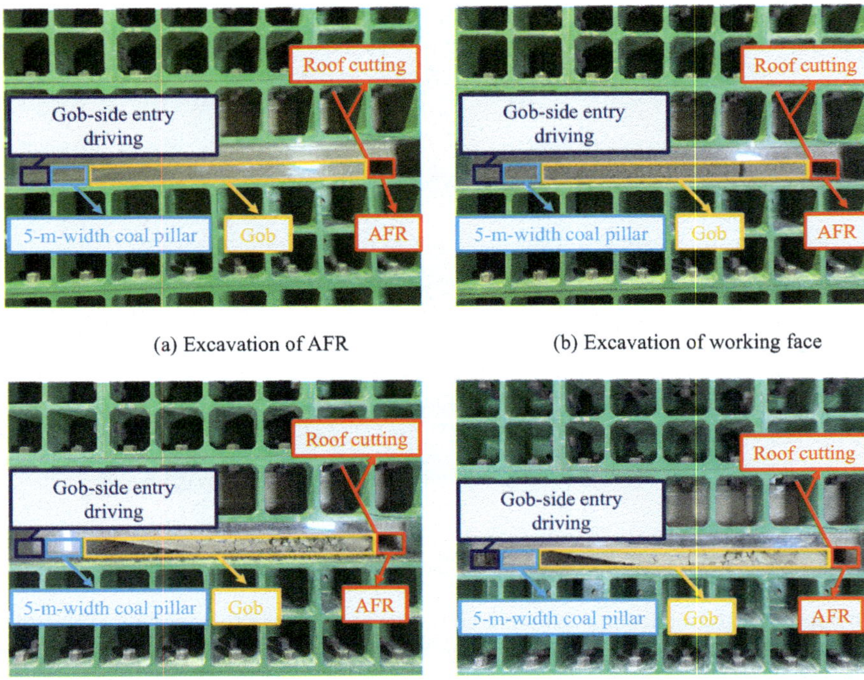

(a) Excavation of AFR (b) Excavation of working face

(c) Formation of AFR (d) Excavation of gob-side entry driving

Fig. 5.17 Excavation process of the model test. Reprinted from Ref. Wang et al. (2021), Copyright © 2021 Published by Elsevier B.V. on behalf of China University of Mining & Technology

5.3.3 Test Process

The specific excavation process of the model test is shown in Fig. 5.17.

5.3.4 Result Analysis

(1) **Analysis of stress distribution of working face**

The contrast of rock stress, roadway convergence defamation, and roof internal displacement are achieved in the AFR and gob-side entry driving based on the analysis of the monitoring data of roadway excavation and overload test as shown in Figs. 5.18 and 5.19.

From Figs. 5.18 and 5.19, the following analysis may be obtained.

1) As the working face mining, the vertical stress of the underlying strata takes on an asymmetrical shape, with low stress in the middle and high stress at either

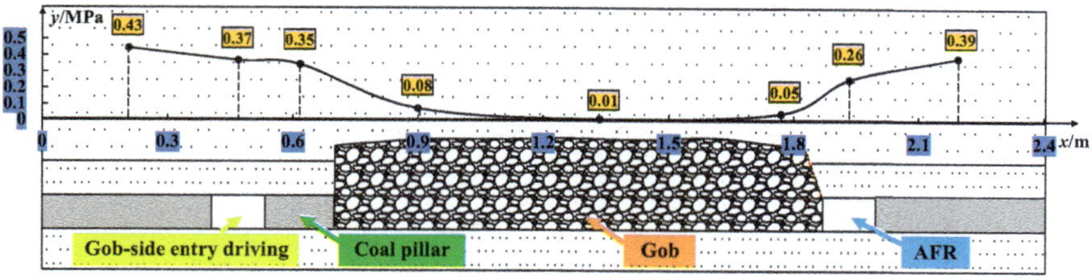

(a) Stress distribution law of overlying strata

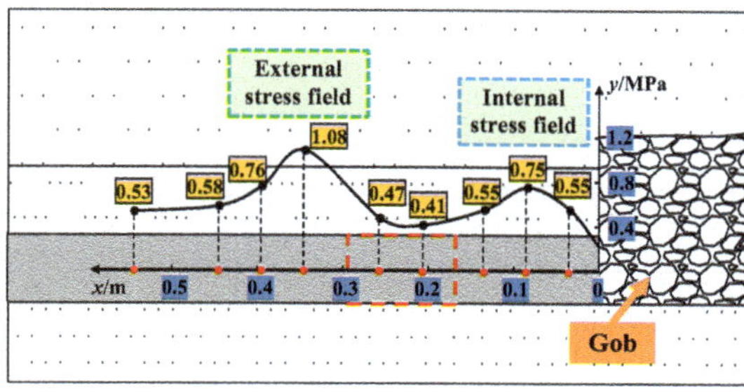

(b) Mining completion

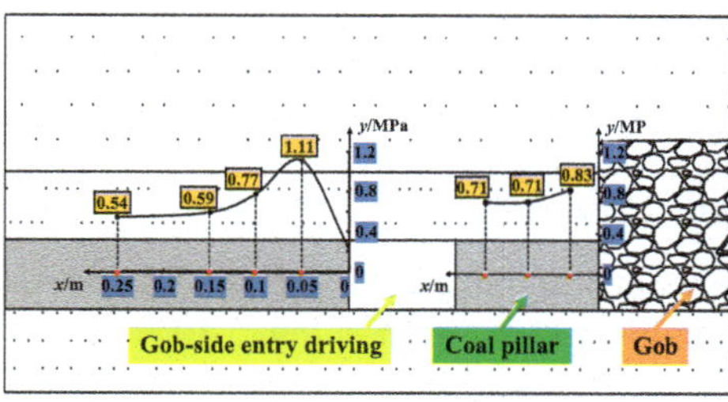

(c) Gob-side entry driving

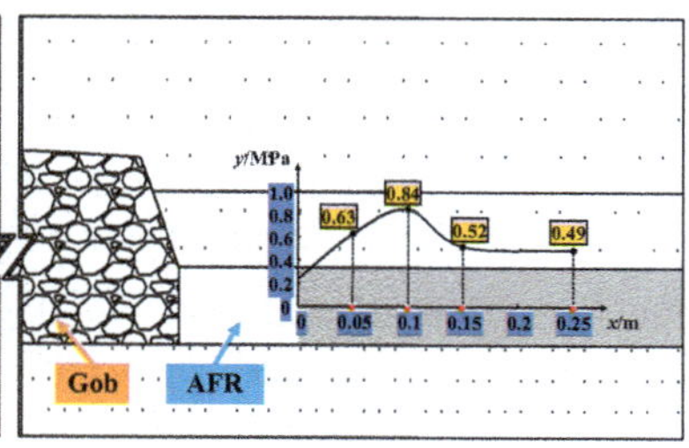

(d) Formation of AFR

Fig. 5.18 Distribution of side abutment pressure. Reprinted from Ref. Wang et al. (2021), Copyright © 2021 Published by Elsevier B.V. on behalf of China University of Mining & Technology

end (Fig. 5.18a). The AFR's overlaying strata have a vertical stress of 0.26–0.39 MPa, but gob-side entry driving has a vertical stress of 0.35–0.43 MPa. At the same location, the former's vertical stress is substantially less than the latter's. This suggests that roof cutting interrupts the stress transmission, which lowers the stress concentration in the roadway surrounding rock.

2) The internal and exterior stress fields prior to the roadway excavation of the gob-side entry driving are presented by the side abutment pressure. The range of 0.15–0.25 m has a low-stress area where the minimum stress is 0.41 MPa. Condensed abutment pressure can be lessened by excavating a roadway in this low-stress area. The mine pillar region is under a lot of stress (Fig. 5.18c). The mine pillar is a key location for the easy collection of surrounding rock energy, which is crucial for managing roadway surrounding rocks.

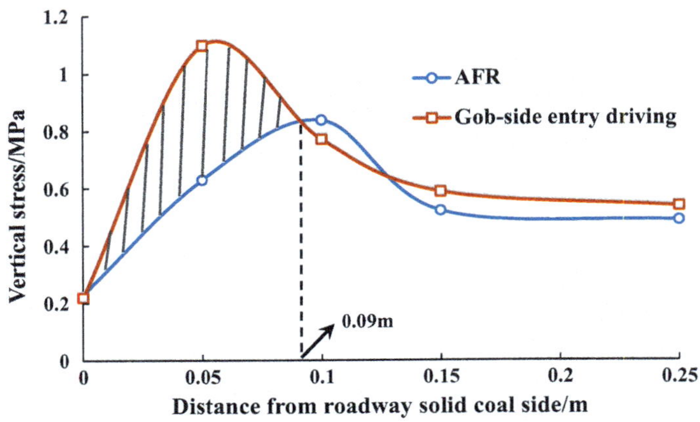

Fig. 5.19 Comparison curves of side abutment pressure. Reprinted from Ref. Wang et al. (2021), Copyright © 2021 Published by Elsevier B.V. on behalf of China University of Mining & Technology

3) As the roadway is formed, side abutment pressure rises and then gradually falls in the direction away from the surface of the solid coal side (Fig. 5.18c, d). AFR and gob-side entry driving have side abutment peak pressure values of 0.84 MPa and 1.11 MPa, respectively. The fact that there is a 24.3% difference in their rates indicates how important AFR's pressure-releasing impact is.

(2) **Analysis of roadway stress monitoring result**

The comparison of surrounding rock stress in different parts of the roadway between AFR mining method and goaf excavation method is shown in Fig. 5.20.

From Fig. 5.20, the following analysis may be obtained.

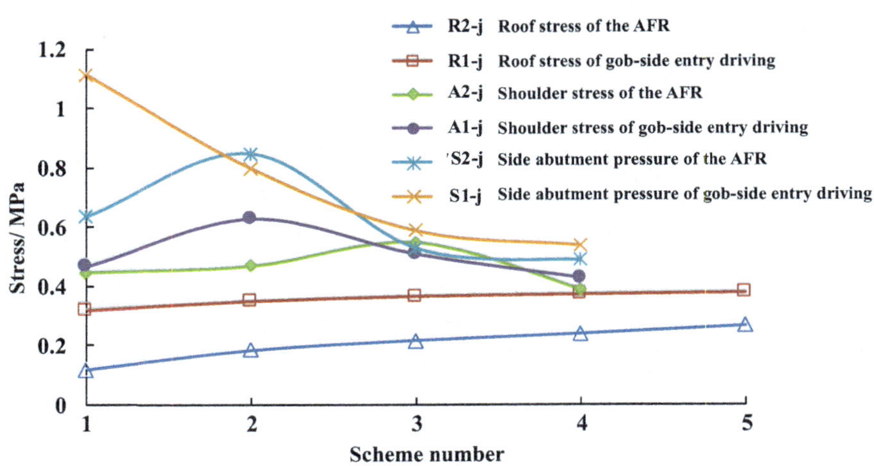

Fig. 5.20 Comparison curves of surrounding rock stress

1) The farther the roof of two types of roadways is from the surface of the roadway, the greater the stress on the roof and gradually tends towards the original rock stress. Within the monitoring range, the maximum stress on the roof of the gob-side entry driving is 0.38 MPa at a distance of 0.25 m (R_{1-5}) from the roadway surface, at a distance of 0.05 m (R_{1-1}) from the roadway surface, the minimum is 0.32 MPa. The maximum stress on the roof of the AFR is 0.27 MPa when it is 0.25 m (R_{2-5}) away from the roadway, and the minimum is 0.12 MPa when it is 0.05 m (R_{2-1}) from the roadway. At the same position of the roof, the stress of the roof at each measuring point of the AFR is lower than that of the gob-side entry driving. This indicates that the directionally roof cutting effectively cuts off the stress transmission between the roadway roof and the gob roof, and the roof pressure relief effect is good.
2) The stress at the shoulder of the gob-side entry driving reaches its peak at a distance of 0.1 m (A_{1-2}) from the roadway surface, with a peak stress of 0.63 MPa. The stress at the shoulder of the AFR with roof cutting reaches its peak at a distance of 0.15 m (A_{2-3}) from the roadway surface, with a peak stress of 0.55 MPa, which is 12.7% lower than the former. This indicates that the peak stress in the shoulder can be reduced and transferred to the deeper part through roof cutting pressure releasing.
3) The roadway side stress reaches its peak at a distance of 0.05 m (S_{1-1}) from surface of solid coal side, which is 1.12 MPa. The roadway side stress of the AFR with roof cutting reaches 0.85 MPa when it is 0.1 m (S_{2-2}) away from the surface of the solid coal side, which is 24.1% lower than the former. It can be reduced and transferred to the deep through the roof cutting pressure relief effect.

Based on the analysis of the above results, it can be concluded that: the directionally roof cutting effectively cuts off the stress transmission between the roadway roof and the gob. The roof stress, shoulder stress, and side abutment pressure of the solid coal side all decrease to varying degrees, and the peak stress of the shoulder and solid coal side is far from the surface of the roadway, making it easy to control the roadway surrounding rock.

(3) **Analysis of roadway surrounding rock deformation**

The monitoring data from the digital sampling system and the grating multipoint displacement measuring system are used to monitor roadway surrounding rock deformation. Figure 5.21 displays the curves of roof and floor convergence deformation. Figure 5.22 displays the curves of convergence deformations of the two sides.

From Figs. 5.21 and 5.22, the following analysis may be obtained.

i) The roadway convergence deformation in both the AFR and gob-side entry driving initially develops slowly, then accelerates, and finally stabilizes in accordance with mining. The overall convergence deformation of AFR is smaller than that of gob-side entry driving during the whole working face mining period. The largest convergence deformation of the roof and floor is 7.6 mm in gob-side entry driving and 4.6 mm in AFR (Fig. 5.21). 1.65 times more than the preceding one. Maximum convergence deformations in AFR and gob-side entry driving are

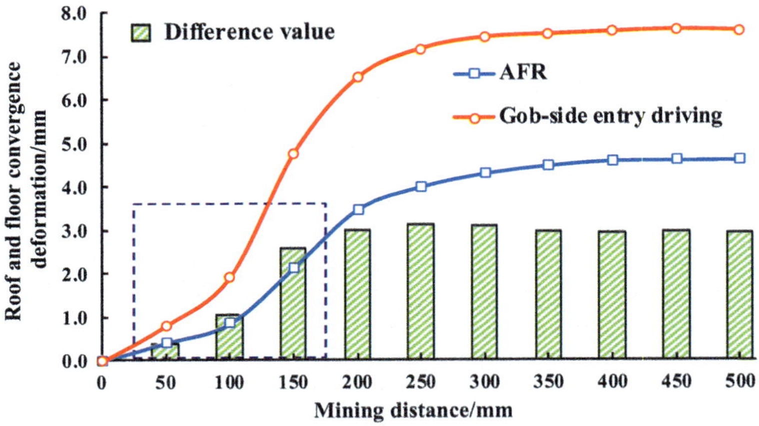

Fig. 5.21 Roof and floor convergence deformation. Reprinted from Ref. Wang et al. (2021), Copyright © 2021 Published by Elsevier B.V. on behalf of China University of Mining & Technology

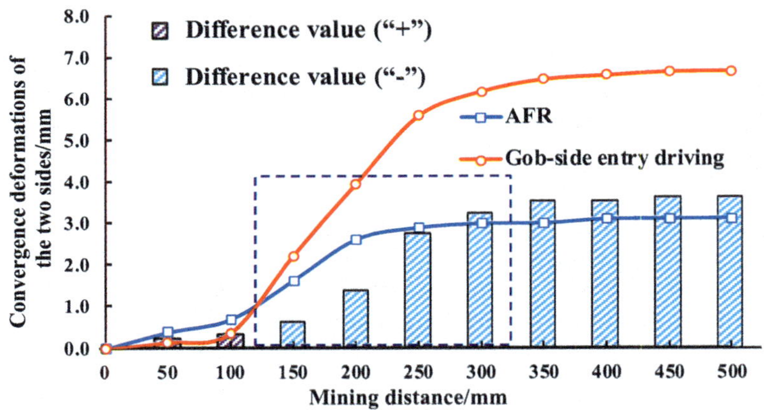

Fig. 5.22 Convergence deformations of the two sides. Reprinted from Ref. Wang et al. (2021), Copyright © 2021 Published by Elsevier B.V. on behalf of China University of Mining & Technology

3.1 mm and 6.7 mm, respectively, for the two sides. The latter is 2.16 times larger than the former (Fig. 5.21). The surrounding rock control effect of AFR mining method is better than that of gob-side entry driving.

ii) With coal mining, the difference value between the gob-side entry driving and AFR's surrounding rock convergence deformation first shows a propensity to rise before eventually tending to remain stable. The larger the differential value, the more control the surrounding rock has. The advantages of the AFR approach in terms of roof and floor convergence deformation become more noticeable

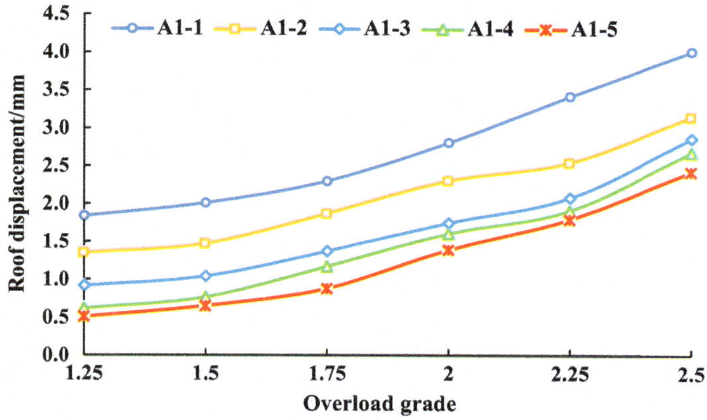

Fig. 5.23 The roof displacement of AFR mining method. Reprinted from Ref. Wang et al. (2021), Copyright © 2021 Published by Elsevier B.V. on behalf of China University of Mining & Technology

when the mining distance increases from 0 to 150 mm (Fig. 5.21). The previously described advantages are more noticeable in terms of the convergence deformations of the two sides when the mining distance increases from 100 to 300 mm (Fig. 5.22). The aforementioned research indicates that the benefits of surrounding rock management on the roadway roof and floor, which are attained by the use of the AFR mining method, are more pronounced than those of the roadway two sides.

(4) **Analysis of overload test**

The internal displacement curves in the AFR roof, obtained during the overload test at monitoring section I ($A_{1\text{-}1}$–$A_{1\text{-}5}$) are shown in Fig. 5.23. The internal displacement curves of the roof for the gob-side entry driving, as measured at monitoring section I ($A_{2\text{-}1}$–$A_{2\text{-}5}$) during the overload test, are shown in Fig. 5.24. Figure 5.25 is comparison of roof internal displacement of two mining methods.

From Figs. 5.23, 5.24 and 5.25, the following analysis may be obtained.

1) In AFR and gob-side entry driving, the roadway roof's internal displacement at shallow monitoring locations is greater than at deep monitoring locations. When the overload grade is higher than 2.0, there is a noticeable acceleration of the roof internal displacements in both types of roadways. In field engineering, the support should now be adequately reinforced to prevent additional deformation of the surrounding rock.
2) In the overload test, the gob-side entry driving at the same roof position had a greater internal displacement than AFR roof. This suggests that the benefits of roof cutting are completely exploited by AFR mining method employed to enhance the surrounding rock's ability to maintain itself and to cut off the stress transfer from the roadway roof. In high-stress situations and mining disturbance

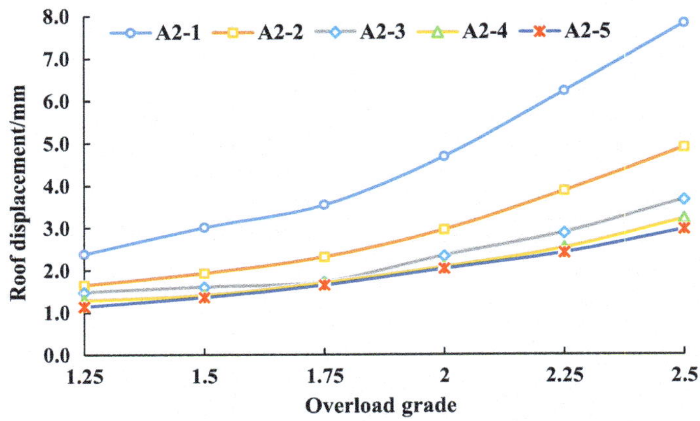

Fig. 5.24 The roof displacement of gob-side entry driving method. Reprinted from Ref. Wang et al. (2021), Copyright © 2021 Published by Elsevier B.V. on behalf of China University of Mining & Technology

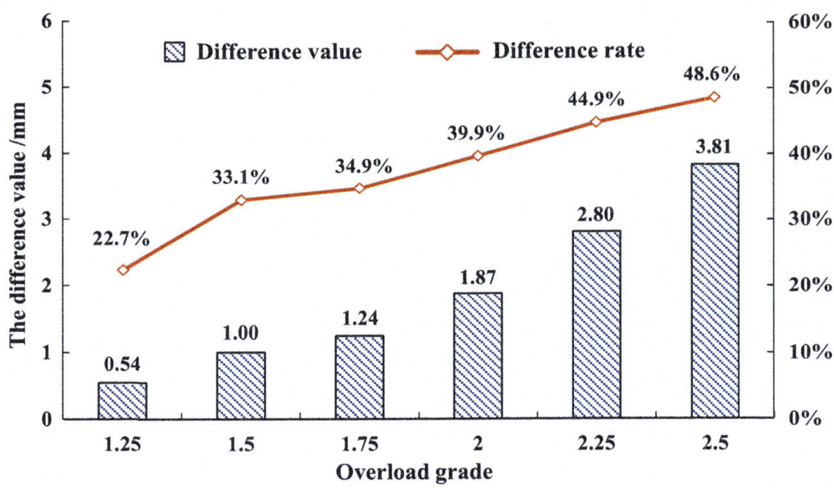

Fig. 5.25 Comparison of roof internal displacement of two mining methods. Reprinted from Ref. Wang et al. (2021), Copyright © 2021 Published by Elsevier B.V. on behalf of China University of Mining & Technology

is present, the AFR surrounding rock control effect may be superior to that of gob-side entry driving due to the aforementioned benefits.

3) As the overload grade increases, the roof internal displacements of the AFR and gob-side entry driving initially increase gradually before accelerating. In terms of both difference value and difference rate, the maximum roof internal displacement in the two types of roadways increase from 0.54 mm to 3.81 mm and

from 22.7% to 48.6%, respectively (Fig. 5.25). As the overload grade rises, the advantages of the AFR surrounding rock control effect over gob-side entry drive become more apparent. This suggests that in greater stress situations, the AFR mining method improves roadway stability more than gob-side entry driving method.

5.4 The AFR in Fault Structural Areas

5.4.1 Simulation Object

In order to confirm the pressure relief impact of the AFR technique in fault structural regions and get the stress evolution law of the roadways and working faces, a three-dimensional geomechanical model test system is utilized to conduct comparative model test study. The study's fundamental characteristics are in line with Sect. 4.3.1, and it is carried out on the Xinjulong Mine's typical fault structure area roadways. The overall situation of the specific model test is shown in the Fig. 5.26.

Fig. 5.26 The model test of the AFR with fault structure. Reprinted from Ref. Wang et al. (2022), Copyright © 2022, The Author(s), under exclusive licence to Springer Nature Switzerland AG

5.4.2 Scheme Design

(1) **Model parameters**

For simulation, the geological range of 440 m × 192 m × 240 m (width × height × thickness) is chosen based on model test requirements and field geological circumstances. The converted model size is 5500 mm × 2400 mm × 3000 mm, and the geometric similarity scale is 80. The bulk density similarity scale is 1.2, and the stress similarity scale is 96.

In order to replicate the real working face with the dip length of 264 m, working face size of model test is 3300 mm. The roadway model, which is used to replicate the real roadway size of 5.1 × 4.2 m, has dimensions of 64 × 53 mm (width × height).

Roof cutting plate of 300 mm is chosen for stress release with an actual roof cutting height of 24 m and a roof cutting angle of 10°. The pressure-releasing segment, which simulates a field roadway length with pressure-releasing of 0–48 m, is located 0–600 mm from the model body's back surface.

(2) **Loading and excavation scheme**

The horizontal load stress of the working face position is 0.47 MPa, and the vertical load stress is 0.25 MPa.

$W_{ri} \rightarrow L_i \rightarrow U_i \rightarrow W_{li}$ is the order in which the excavation is carried out. Detailed excavation schemes are displayed in Fig. 5.27.

1) Working face is mined in 30 steps, with 100 mm of excavation footage for each step of W_{r1}–W_{r30}.
2) There are thirty excavation steps of L_1–L_{30} for the 2305S Haulage roadway, with a 100 mm excavation footage for each step.
3) There are thirty steps in the excavation of 2305S Tail roadway, with 100 mm excavation footage for each step of U_1–U_{30}.
4) Excavation of working face takes place in 60 steps, measuring 50 mm for each step in W_{l1}–W_{l60}.

The excavation of the 2305S working face is stopped once it reaches the 48th step, and pressure relief methods utilizing directional roof cutting with a deep hole are applied to the gob-side roof of the 2305S tail roadway. The roof cutting plate that has been implanted beforehand must be gradually removed in steps of 50 mm. The next 12 steps will continue the excavation of the 2305S working face after the roof cutting plate has been fully removed.

(3) **Monitoring scheme**

Two monitoring sections, I and II, are set up along the strike direction of the simulated working face to conduct a comparison study on the disaster brought on by faults in the roadway. Another monitoring section III is positioned 300 mm from the model body's rear surface (0–600 mm distant from the back surface of the model body). Every monitoring point inside every monitoring segment undergoes stress monitoring. The

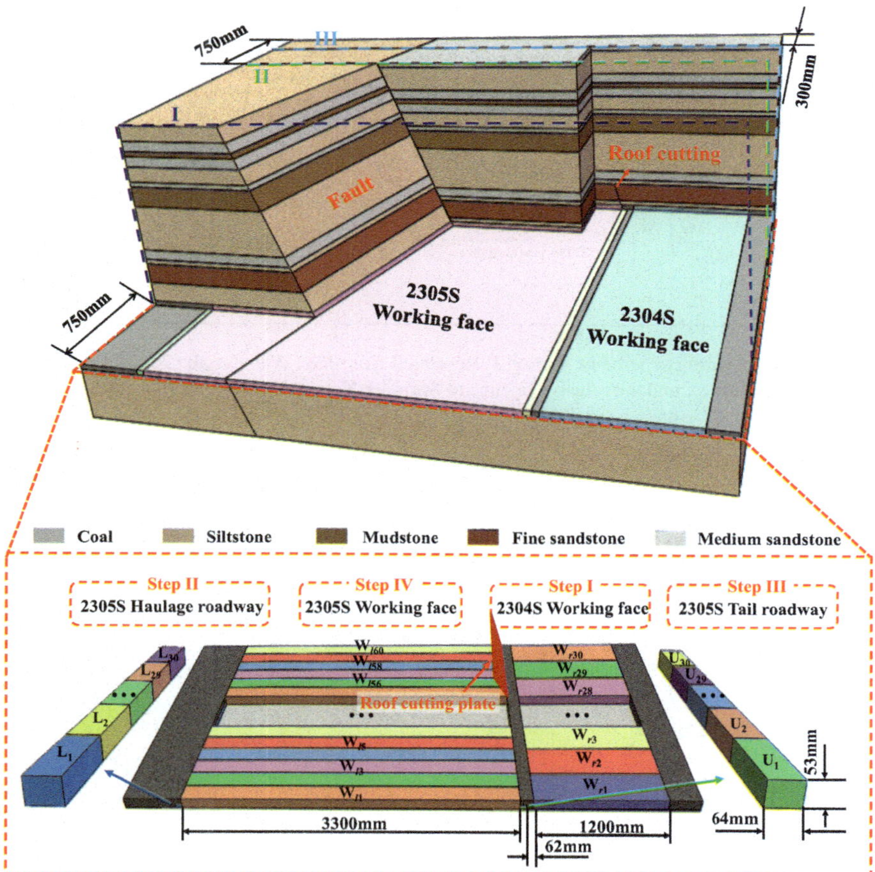

Fig. 5.27 The excavation scheme. Reprinted from Ref. Wang et al. (2022), Copyright © 2022, The Author(s), under exclusive licence to Springer Nature Switzerland AG

monitoring points' exact positions are as follows: *i* stands for the part on monitoring, i = I–III; j denotes the monitoring point. There are R_{i-j}, S_{i-j}, and W_{i-j} number of monitoring locations for the working face, shoulder, and roof of the roadway. Figure 5.28 shows the arrangement of monitoring points using monitoring section I as an example.

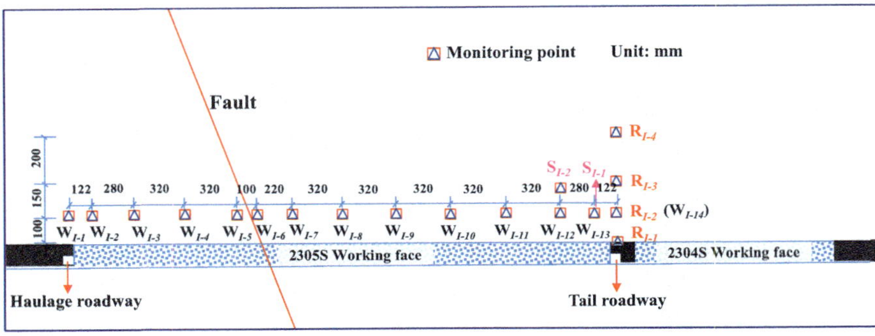

Fig. 5.28 Monitoring scheme of section I. Reprinted from Ref. Wang et al. (2022), Copyright © 2022, The Author(s), under exclusive licence to Springer Nature Switzerland AG

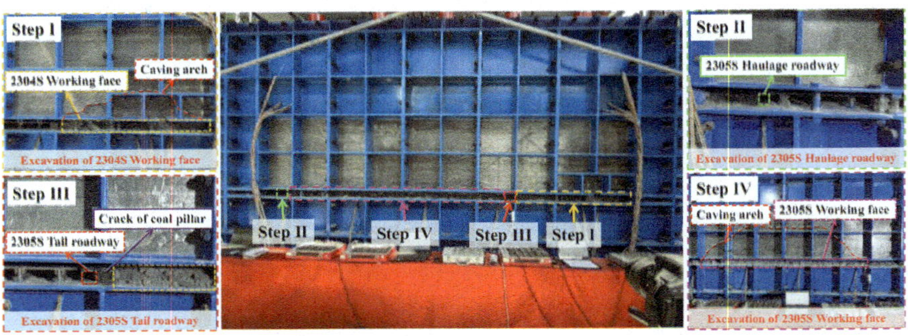

Fig. 5.29 The model test process. Reprinted from Ref. Wang et al. (2022), Copyright © 2022, The Author(s), under exclusive licence to Springer Nature Switzerland AG

5.4.3 *Test Process*

The specific excavation process of the model test is shown in Fig. 5.29.

5.4.4 *Result Analysis*

(1) **Analysis of the working face stress**

For every monitoring point above the working face roof selected as the research object, the stress distribution law of the working face mining through the monitoring sections I, II, and III is assessed, as shown in Fig. 5.30.

From Fig. 5.30, the following analysis may be obtained.

1) Comparison of working face roof stress characteristics: Together with the mining of the 2305S working face, the roof in front of the working face has

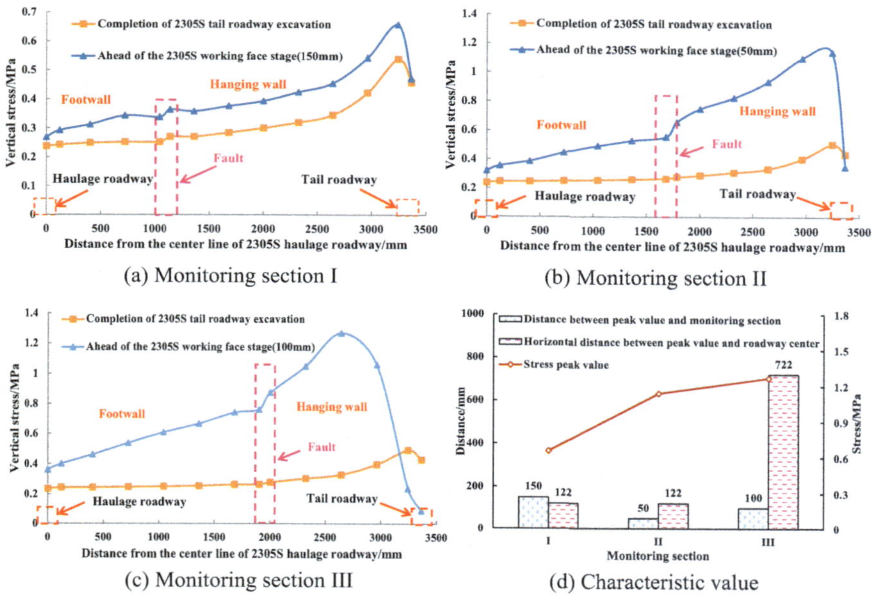

(a) Monitoring section I

(b) Monitoring section II

(c) Monitoring section III

(d) Characteristic value

Fig. 5.30 The working face roof stress analysis. Reprinted from Ref. Wang et al. (2022), Copyright © 2022, The Author(s), under exclusive licence to Springer Nature Switzerland AG

progressed influence pressure. Additionally, following the 2305S tail roadway excavation, it is superimposed with the distribution pressure of the surrounding rock, causing the stress of the working face roof surrounding rock to increase even more.

2) Comparative of fault influence degree: The stress peak of the working face roof rises from 2.64 times of the starting rock stress to 4.56 times as the working face is excavated from monitoring section I to monitoring section II. It is becoming more evident how advance pressure impacts the working face roof and surrounding rock that the stress peak is 50 mm ahead of the working face instead of 150 mm. Meanwhile, the average stress on the working face roof increases to 0.82 MPa where the fault hanging wall is located, representing an increase of 82.2% over the prior value of 0.45 MPa.
3) Comparative of pressure release effect: The working face roof's stress peak point is located 722 mm from the center of the roadway, a greater distance when the working face is excavated from monitoring section II to monitoring section III. By moving the stress peak location forward of the working face from 50 to 100 mm, the focused stress for the roadway has less of an impact on the surrounding rock of the working face roof. Meanwhile, the working face roof's average stress, which is 7.3% lower than the previous, drops from 0.82 MPa to 0.76 MPa where fault hanging wall is situated.

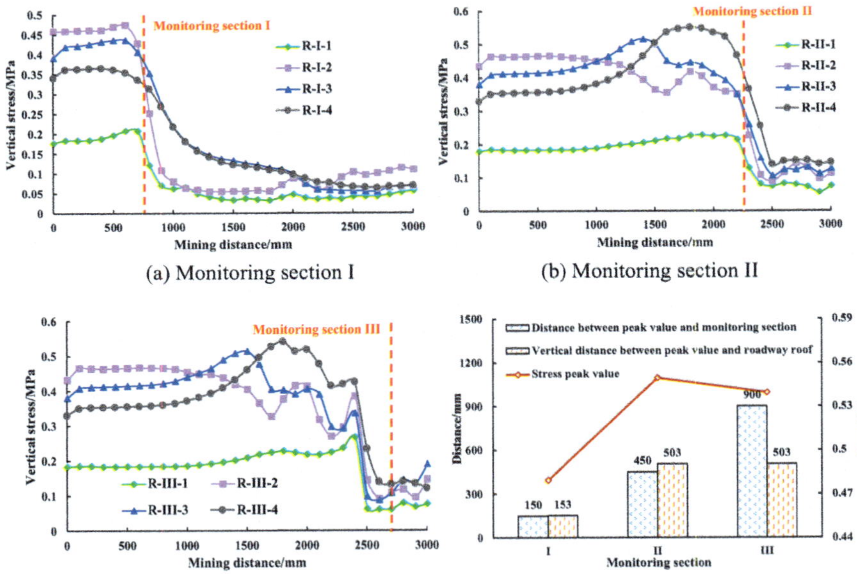

(a) Monitoring section I

(b) Monitoring section II

(c) Monitoring section III

(d) Characteristic value

Fig. 5.31 The roadway roof stress. Reprinted from Ref. Wang et al. (2022), Copyright © 2022, The Author(s), under exclusive licence to Springer Nature Switzerland AG

(2) **Analysis of the roadway roof stress**

Figure 5.31 displays the stress evolution laws of the roadway roof through the monitoring sections I, II, and III.

From Fig. 5.31, the following analysis may be obtained.

1) Comparison of the stress properties of the roadway roof surrounding rock: The monitoring section I exhibits a constant rise in surrounding rock stress throughout the region of 0–700 mm, suggesting that the advanced effect range extends beyond 700 mm. When the working face is excavated 700 mm or the monitoring section II is positioned 1550 mm forward of the working face, there is a noticeable shift in the surrounding rock stress of the roadway roof. It demonstrates that the influence range of the roof advance stress is from 0 to 1550 mm forward of the working face.
2) Comparison of the degree of fault influence: The advance stress of the roadway roof peaks at 150 mm forward of the working face, which is the position of monitoring section I. The stress peak is 0.48 MPa and is situated at monitoring point $R_{I\text{-}2}$ of the roadway roof. It is 1.92 times the initial rock stress. When working face mining is used to monitor section II, the roadway roof's advance stress peak first appears at monitoring point $R_{II\text{-}2}$. The stress peak exhibits continuing mining of the working face and continuous transmission to the deep monitoring stations of the surrounding rock roof. The maximum roof advance stress peak is

reached when the monitoring section is positioned 550 mm ahead of the working face. At monitoring point R_{II-4} on the roadway roof, the stress peak is apparent at 0.55 MPa, or 2.2 times the initial rock stress.

3) The advance stress peak drops to 0.54 MPa when the working face is excavated from monitoring section II to III. The stress peak is positioned 900 mm forward of working face instead of 450 mm, and the advance pressure's effect on the roadway roof surrounding rock is lessened. The findings indicate that the roof directional cutting significantly contributes to roof pressure-releasing control in closer proximity to the fault area.

(3) **Analysis of the roadway shoulder stress**

Figure 5.32 displays the stress evolution laws of the roadway left shoulder through the monitoring sections I, II, and III.

From Fig. 5.32, the following analysis may be obtained.

1) Comparison of the degree of fault influence: The monitoring section I is located 150 mm forward of the working face, where the advance stress of the roadway shoulder peaks. At 0.66 MPa, the stress peak is 2.64 times greater than the initial rock stress. Advance stress of roadway shoulder peaks at 50 mm forward of the working face, which is the location of monitoring section II. At 1.14 MPa, the stress peak is 4.56 times greater than the initial rock stress. In the two monitoring

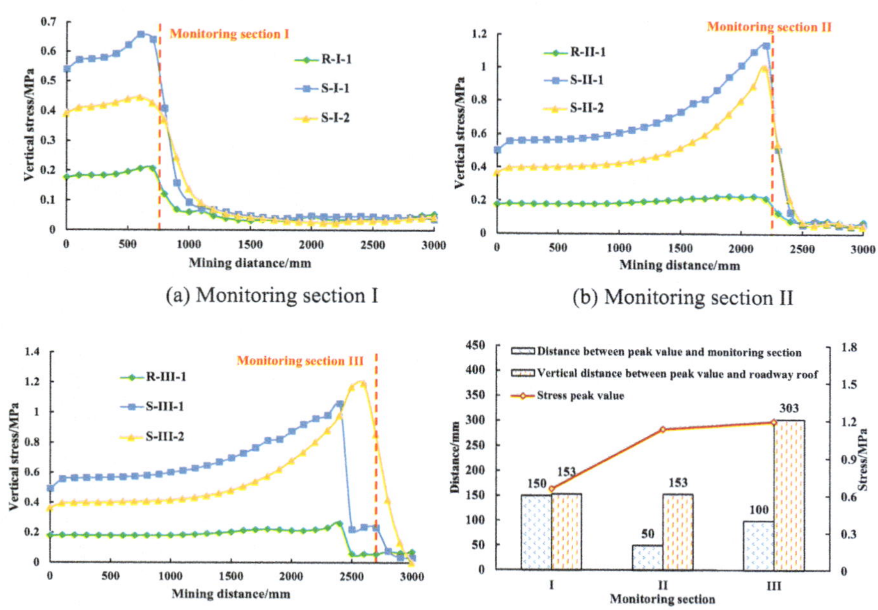

(a) Monitoring section I (b) Monitoring section II

(c) Monitoring section III (d) Characteristic value

Fig. 5.32 The roadway left shoulder stress. Reprinted from Ref. Wang et al. (2022), Copyright © 2022, The Author(s), under exclusive licence to Springer Nature Switzerland AG portions, there is a vertical space of 153 mm between the roadway shoulder's stress peak point and the roadway roof surface.

2) Comparison of methods for relieving pressure: The advance stress peak point of the roadway shoulder and the roadway roof surface are separated by 303 mm instead of 153 mm when the working face is excavated from monitoring section II to III. By moving the stress peak location forward of the working face from 50 to 100 mm, the concentrated stress for the roadway has less of an impact on the shoulder surrounding rock. The impact of advance pressure on the shoulder of the roadway surrounding rock lessens.

5.5 Chapter Summary

The model test system of AFR without mine pillars with an integrated mining and roadway retention mode is developed. Geomechanical model tests have been conducted on different types of working faces, such as shallow buried depth, deep buried high stress, and fault structural areas. The control mechanism of AFR under different condition parameters is clarified.

References

Wang Q, Qin Q, Jiang B, Jiang ZH, He MC, Li SC, Wang Y (2020) Geomechanics model test research on automatically formed roadway by roof cutting and pressure releasing. Int J Rock Mech Min Sci 135:104506

Wang Q, He MC, Li SC, Jiang ZH, Wang Y, Qin Q, Jiang B (2021) Comparative study of model tests on automatically formed roadway and gob-side entry driving in deep coal mines. Int J Min Sci Technol 31(4):591–601

Wang Q, Wang Y, He MC, Li SC, Jiang ZH, Jiang B, Xu S, Wei·HY (2022) Experimental study on the mechanism of pressure releasing control in deep coal mine roadways located in faulted zone. Geomech Geophy Geo-Energy Geo-Resources 8(2):1–24

Chapter 6
Engineering Application of AFR Mining Method

The specific design method of AFR mining method will be proposed in this chapter. Meanwhile, the application research will carry out in different types of mine working faces including shallow buried depth, deep buried depth, weathered roof strata, broken roof strata, near the overlying mined coal seam and other typical examples. The specific application situation will be introduced in the following section.

6.1 Design Principle of Key Parameters

Based on the previous theoretical, numerical, and model test research, the design principle for the roof cutting parameters is obtained, as follows.

6.1.1 Roof Cutting Parameters Design

(1) **Roof cutting height design**

Due to the expansion characteristics, the collapsed rock fragments fill the gob. Based on the expansion theory in the rock mass (Wang et al. 2022), the roof cutting height can be calculated by

$$H \geq (M - \Delta H_1 - \Delta H_2)/(K - 1) \tag{6.1}$$

where H is the cutting height, m; M is the mining height, m; ΔH_1 is the deformation of the roof, m; ΔH_2 is the floor heave, m; and K is the broken expansion coefficient of the rock mass. K varies as a function of field condition and can be determined by the broken expansion equation obtained from field measurements.

Q. Wang et al., *Theory and Method of Automatically Formed Roadway in Underground Mining*, https://doi.org/10.1007/978-981-97-8082-2_6

Table 6.1 Roof cutting angle ranges

Type	Thin coal seams ($t <$ 1.3 m)	Medium-thickness coal seams (1.3 m $\leq t \leq$ 3.5 m)	Thick coal seams ($t >$ 3.5 m)
Cutting angle/°	15–20	15–20	10–15

The stress relief effect from the roof rock increases with increasing cutting height *H*. When the cutting height exceeds a certain value, the stress relief effect is no longer noticeable. Therefore, it is important to consider both construction cost and construction efficiency when determining the cutting height.

(2) **Roof cutting angle design**

A reasonable cutting angle can reduce the disturbance from the blasting to the support system and the friction at the cutting interface when the roof collapses. Too small a cutting angle will cause a large frictional resistance to the roof, while blasting will cause damage to the cables near the roof that is being cut. Conversely, too large a cutting angle will increase the length of the roof beam and lead to poor backfilling effect in the gob below the cutting roof.

As a result, determining the cutting angle of the roof should consider the coal seam thickness. Previous studies have shown that the reasonable roof cutting angle is from 10° to 20° (Wang et al. 2022). Table 6.1 lists the roof cutting angle ranges with different coal seam thicknesses.

6.1.2 *Support Parameters Design*

Through the previous series of studies, it can be seen that the cable and single prop (support) are the main support components in AFR. The cable is a permanent support installed during roadway excavation. The single prop (support) is a temporary support installed during the mining period of the working face. To ensure the safety and stability of AFR, it is necessary to comprehensively consider the design of cable and single prop (support).

(1) **Design of cable support parameters**

In terms of cable length design, as a prestressed support component, cables need to be anchored to the complete rock stratum so as to exert their effectiveness. Therefore, the design length of the cables need meet the requirement that their anchoring section is anchored to a complete rock stratum of more than 1.0 m. Meanwhile, the cables should ensure the stability of the short cantilever beam formed by roof cutting, and the length of the cables should be at least 2.0 m greater than the roof cutting height. Determine the design length of cables to meet both of the above conditions simultaneously.

In terms of cable prestress design, traditional cables can be prestressed to 30–50% of their yield load. The energy absorbing cables have large deformation characteristics and high safety reserve after applying high prestress. Therefore, the prestress can be designed as 70–90% of their yield load.

In terms of cable support spacing design, according to the design equation for AFR support force of different zone mentioned in Sect. 2.4 (Eq. 2.33), determine the required support force F per unit area of AFR surrounding rock in different zones. Based on the minimum F value in each zone, a reasonable support spacing parameter is design by combining with the prestress of cables. In this way, the zone with the smallest required support force F in different zones can only meet safety requirements with anchor cables. For the other two zones with larger required support force F, temporary support (single prop/support) should be used.

(2) **Design of single prop (support) parameters**

The single prop (support) in the AFR mainly contains traditional hydraulic prop (support) and the new equipment which includes confined concrete props and the roof & rib support equipment in this book. The work resistances of traditional hydraulic prop (support) will be set during production. The work resistances of confined concrete props and the roof & rib support equipment are described in Sect. 3.4.

In the other two zones with larger required support force F, it is necessary to install single prop (support) on the basis of cables support. Subtract the smallest F value of the zone which only used cables from the larger F value of two zones, and then calculate the support spacing of the single prop (support) to ensure the safety of the AFR in the entire mining process.

6.2 The AFR with Shallow Buried Depth

6.2.1 Engineering Object

The engineering geology and application conditions of the AFR mining method with shallow buried depth in the Ningtiaota Mine is consistent with Sect. 4.1.1. This innovative mining technique splits the roof above the gob (Part B) and above the roadway (Part C) using directionally roof cutting. Due to gravity and mine pressure, the roof above the gob falls into the gob, causing the broken rock mass fragments to further expand and fill the gob (Part D). Thus, the AFR (Part E), which will be utilized for the next working faces, is automatically formed and mine pillars are not required. The specific on-site construction process is shown in Fig. 6.1.

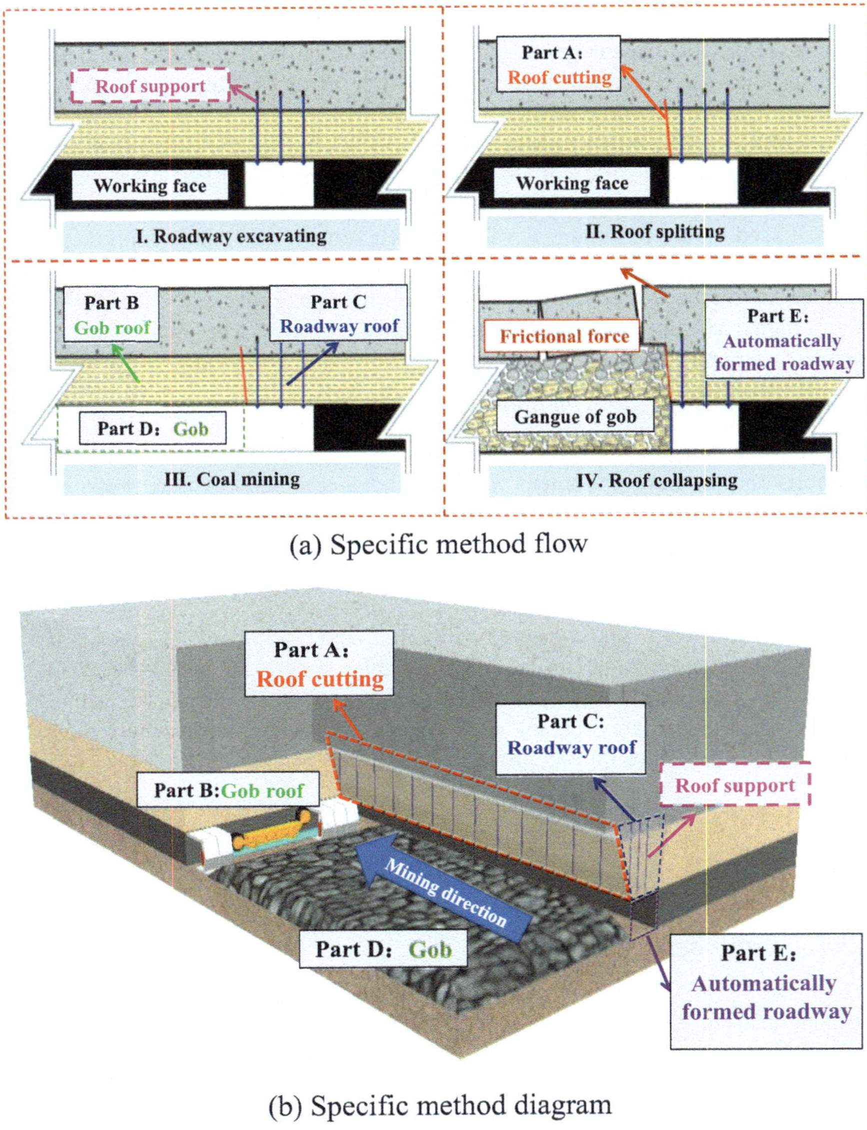

Fig. 6.1 Application of the AFR mining method with shallow buried depth

6.2.2 Scheme Design

Figure 6.2 displays the design parameters for AFR by roof cutting pressure release without mine pillars. (1) Roof cutting parameters: The cutting angle is 15° and the cutting height is 9.0 m. (2) Roof support parameters: The size of high-prestress

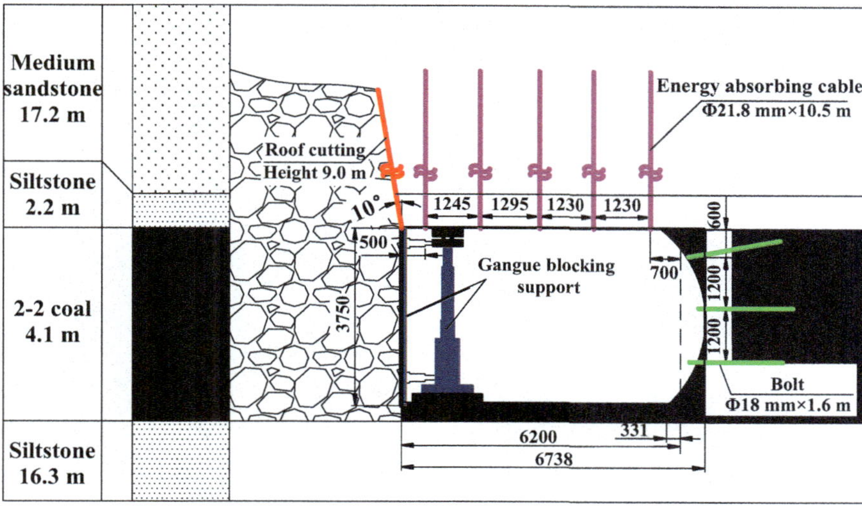

Fig. 6.2 Design diagram of roadway roof cutting and support parameters in Ningtiaota Mine. Reprinted from Ref. Wang et al. (2020a, b), Copyright © 2020 Elsevier Ltd.

energy absorbing cables is Φ21.8 mm × 10.5 m for supporting the roadway roof. (3) Gangue-blocking support parameters: "Roof and rib support equipment + gangue retaining steel mesh + expandable U-shaped steel" is the gangue blocking support utilized on the gob side of the AFR.

6.2.3 Application and Monitoring Analysis

The AFR mining method is suggested for the Ningtiaota Mine's S1201-II working face. Install displacement measurement stations I, II, and III along the roadway behind the working face to track the deformation of the roadway caused by the convergence deformation of the roof and floor during mining operations. Figure 6.3 depicts the on-site AFR effect at formation stage and reuse stage of the AFR. Figure 6.4 shows the variation curve of the AFR's roof and floor convergence deformation at the typical measurement station.

The following conclusions may be drawn from the monitoring of the curve for the roof and floor convergence deformation at the typical measurement station. The variation law of roadway's roof and floor convergence deformation shows the following four obvious periods. (1) The pressure relief impact stage: the working face is lagged from 0 to 40 m and the amount of roof and floor convergence deformation is small. (2) The dynamic pressure deformation rising stage: the working face is lagged from 40 to 95 m, the roof and floor convergence deformation rapidly increases. (3) The dynamic pressure stabilizing stage: the working face is lagged from 95 to

(a) Formation stage of the AFR

(b) Reuse stage of the AFR

Fig. 6.3 The field application effect of the AFR mining method in Ningtiaota Mine. Reprinted from Ref. Wang et al. (2020a, b), Copyright © 2020 Elsevier Ltd.

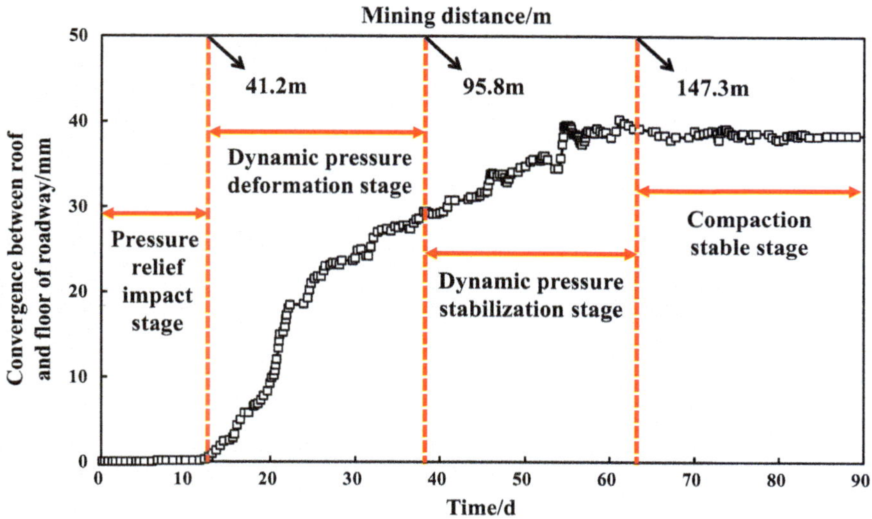

Fig. 6.4 Roof and floor convergence deformation at the typical measurement station. Reprinted from Ref. Wang et al. (2020a, b), Copyright © 2020 Elsevier Ltd.

145 m, the roof and floor convergence deformation slows down and tends to stabilize. (4) The compaction stable stage: after the working face is lagged about 145 m, the deformation of surrounding rocks basically tends to stabilize.

The above monitoring data indicates that the AFR mining method effectively controls the safety and stability of roadway with shallow buried depth.

6.3 The AFR of Weathered Roof Strata

6.3.1 Engineering Object

The Hongliulin Mine is located in western China. The 44,203 working face is in thick weathered strata (maximum thickness: 130 m) with a thin bed rock strata (minimum thickness: 27 m). The size of the roadway is 6000 mm × 3400 mm, with a coal seam thickness of 3.6 m. The immediate roof is composed of 8.1 m fine sandstone, the basic roof is 20.2 m medium sandstone, and the immediate roof is 2.7 m sandy mudstone.

The mine pressure appears violent due to the overlying strata above the working face. This provides a huge challenge for controlling the surrounding rock of the roadway. This working face is mined using the AFR mining method, which can reduce roadway excavations and eliminate mine pillar.

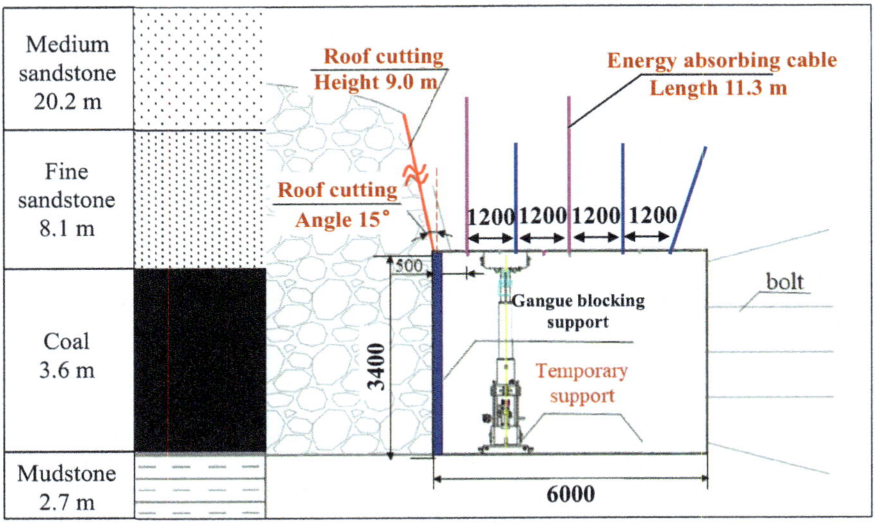

Fig. 6.5 Design diagram of roadway roof cutting and support parameters in Hongliulin Mine

6.3.2 Scheme Design

The key to this mining method is to cut off the physical connection between the roof of the gob and the roof of the roadway. At the same time, high-strength energy absorbing support is required to ensure the stability of the surrounding rock of the AFR under the combined influence of the friction caused by the collapse of the gob roof and the mining disturbance of the working face. The energy absorbing anchor cables is used to support the 44,203 working face's belt roadway so as to reduce roof deformation caused by mining. The design parameters of AFR mining method are shown in Fig. 6.5. Specifically, the energy absorbing cables are intended to have a prestress of 190 kN, and the size of the cable is Φ 22 × 11,300 mm. The directional cutting height in AFR is designed to be 9.0 m, and the cutting angle is designed to be 15°.

6.3.3 Application and Monitoring Analysis

The application effect of AFR mining method is shown in Fig. 6.6a. In Fig. 6.6b, the force variation curve of energy absorbing cable, roof and floor convergence deformation using AFR mining method are plotted.

From Fig. 6.6, the following analysis may be obtained.

The collapse of the gob's roof causes the roof and floor convergence deformation of the AFR to rise rapidly from the standpoint of managing rock deformation. After

(a) The application effect at roadway formation 50 m

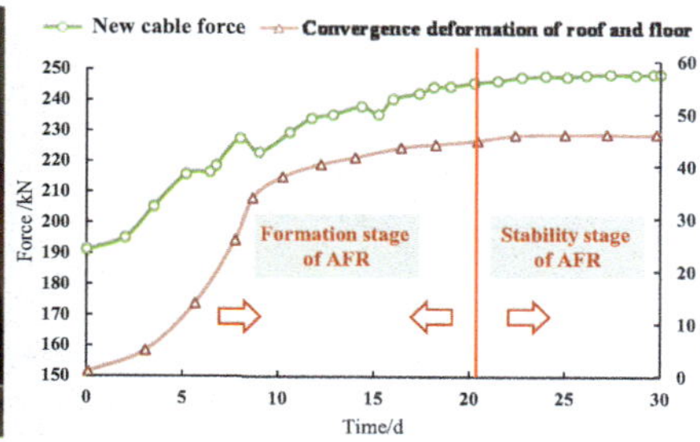

(b) Monitoring curve

(c) The application effect at roadway formation 250 m

Fig. 6.6 Application effect of the AFR of weathered roof strata in Hongliulin Mine

the gob is filled with broken rock mass, the roadway progressively stabilizes, reaching a maximum roof and floor convergence deformation of 46.2 mm. This deformation meets the safety production requirements on site. These findings suggest that the AFR mining method including roof cutting and energy absorbing cables might significantly enhance rock stability in roadways affected by high dynamic disturbances and guarantee safe operating conditions.

The above monitoring data indicates that the AFR mining method effectively controls the safety and stability of roadway with weathered roof strata.

6.4 The AFR of Broken Roof Strata

6.4.1 Engineering Object

The engineering geology and application conditions of the AFR mining method of broken roof strata in the Suncun Mine is consistent with Sect. 4.2.1. The AFR mining method by roof cutting and high-strength bolt-grouting is utilized in the transportation roadway of the working face, so as to prevent large deformation and roof collapse of the roadway and preserve mine pillar resources, as shown in Fig. 6.7.

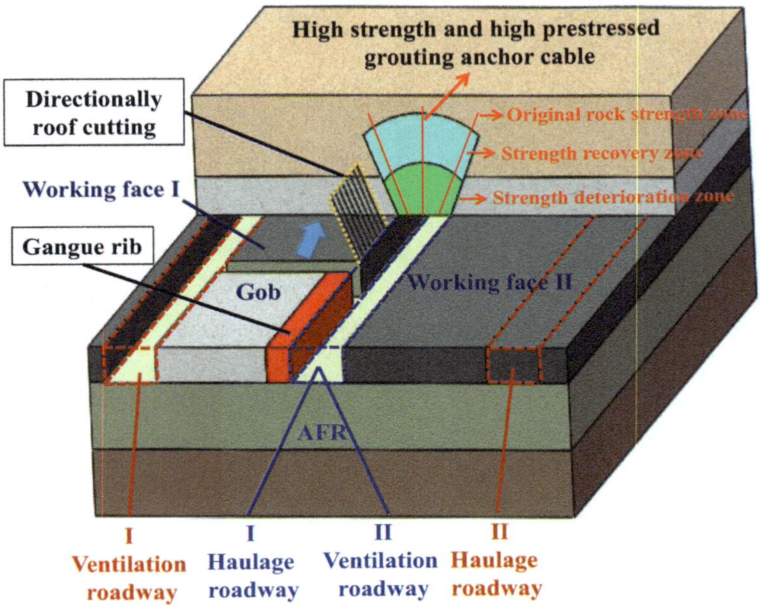

Fig. 6.7 The diagram of AFR mining method by roof cutting and bolt-grouting

The AFR mining method suitable for broken roof mainly includes roof cutting and bolt-grouting. On the one hand, the integrity of the roadway roof is enhanced by the employment of high strength bolt-grouting. However, the directionally roof cutting eliminates stress transmission between the roadway roof and the gob roof, placing the roadway in the stress reduction area.

Specific technical principles and processes are as follows.

(1) High prestressed bolt-grouting cable is used to support roadway roof in mining area. In order to maximize the combined supporting effects of the bolt and the anchor cable as well as the surrounding rock's ability to support itself, on the one hand, high pressure grouting is utilized to fill in the internal cracks of the surrounding rock across a wide range. It also enhances the roof integrity, and improves the surrounding bolt's anchoring performance. On the other hand, the grouting cable is anchored to the strength zone of the original rock, giving full play to its advantages of high strength and high prestress. There is a large compressive stress in the reinforced surrounding rock, and it effectively improves the overall strength of the broken grouting rock mass.
(2) To safeguard the roof along the gob side and the solid coal side of the roadway, hydraulic support is installed. When the roof collapsing creates extreme friction on the roof cutting surface during the AFR, the roof support is reinforced by high-strength bolt-grouting to provide a rigid roof effect, ensuring the safety and stability of the roadway roof.

In addition, automatic roadway forming techniques such as directional roof cutting pressure relief, roadway gangue rib support are consistent with the AFR mining method mentioned earlier.

6.4.2 Scheme Design

Due to the particularity of the broken roof of the roadway, before applying the AFR mining method, it is necessary to detect and analyze the broken degree of the roadway roof surrounding rock. The specific detection analysis is as follows. Eight measurement locations, labelled Nos. #1–#8, are drilled at a depth of 6 m, and boreholes monitoring is conducted on the roadway roof. The design of grouting support parameters is carried out according to the roof surrounding rock detection.

The findings of the drill detection reveals that the immediate roof of the roadway is composed of siltstone, with a thickness that varied from 1.8 to 3 m and roughly reduced in the direction of the working face's mining. The internal breakage on the roadway roof varies in degree and also reduces in the working face's mining direction. As shown in Fig. 6.8, the roadway is separated into three zones based on the roof fracture conditions: Zone I (intact roadway area) is beyond 5.2 m, Zone II (intact roadway area) is beyond 4.8 m, and Zone III (intact roadway area) is beyond 4.5 m.

The roadway roof zones are taken into consideration when designing each zone's control scheme. An example of Zone I, which has the widest damaged range, that is 5.2 m. In order to anchor the grouting anchor cable to a intact rock strata of more than 1.0 m, the length of the grouting anchor cable should be at least 6.2 m. The roof high-strength grouting anchor cables measuring Φ22 mm × 6200 mm (prestress: 120 kN). The roadway is held up by side-to-side bolts measuring Φ20 mm × 2000 mm (inter-row spacing: 1000 mm × 800 mm). 50 m in advance of the working face is the

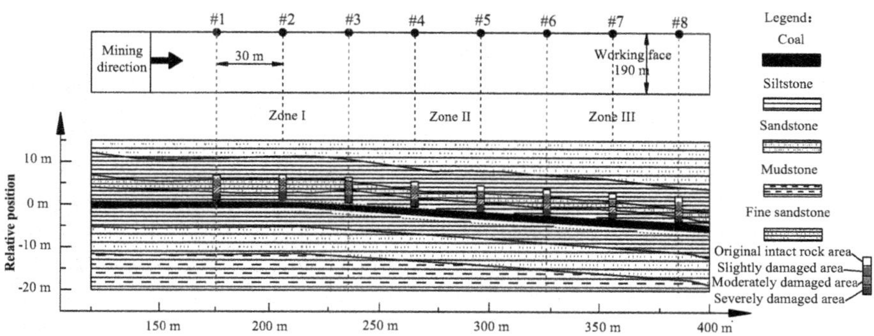

Fig. 6.8 Roadway roof zoning along entire roadway. Reprinted from Ref. Wang et al. (2020a, b), Copyright © 2020 Elsevier Ltd.

range in which pre-splitting and roof cutting are completed. The roof is first supported by the hydraulic props. The roof cutting hole spacing is 800 mm, the hydraulic prop spacing is 2000 mm, the roof cutting height is 3 m, and the roof cutting angle is 15°. In order to sustain the gangue rib 20 m and more behind the working face, gangue rib support by hydraulic support is employed. Figure 6.9 shows the scheme design for the AFR mining method by roof cutting and bolt-grouting.

6.4.3 Application and Monitoring Analysis

On-site hydraulic prop stress, grouting anchor cable stress, and roadway roof & floor convergence deformation are monitored on the test roadway. As shown in Fig. 6.9, the anchor cable dynamometers and hydraulic prop pressure gauges are set up. Figures 6.10, 6.11 and 6.12 depict the surrounding rock deformation and the forces of the supporting members. Figure 6.13 shows the AFR application effect.

From Figs. 6.10, 6.11, 6.12 and 6.13, the following analysis may be obtained.

(1) The results of the roadway deformation monitoring shows that the roadway deformation increases steadily as the working face advanced within the advanced support range. The roadway deformation rate increases between 0 and 40 m behind the working face. The two sides, floor, and roof deformations stabilizes at a position of 60 m behind the working face, with only minor deformation. The magnitudes of deformation decrease in the following order: roof displacement > coal-side displacement > floor displacement > gob-side displacement. The biggest displacement (181 mm) is found on the roof, which satisfies field safety.

(2) The hydraulic prop stress increases from 11.0 to 29.8 kN as the working face advanced within the advanced support range, according to a mechanical analysis of the props. The gangue ribs are essentially compacted and the hydraulic prop stress stabilizes at a position 63 m behind the working face. With a strength utilization ratio of 50.2%, the hydraulic prop on-site has a maximum stress of 125.5 kN and a maximum working resistance of 250.0 kN. These values indicate a suitable stress and an adequate safety margin. In terms of economy, the hydraulic prop spacing of 2000 mm improves the working face construction efficiency in the field.

(3) A study of the variations in the stresses on the three grouting anchor cables reveals that, within the working face's advanced support range, the stresses on the cables rise with the advancement of the working face. The roof cutting collapsed rock mass causes a dramatic drop in grouting anchor cable in the 9–24 m area behind the working face. The grouting anchor cable stresses normally stabilized at a point 24 m behind the working face. The gob-side roof cable stress, coal-mass side cable stress, and mid-roadway roof cable stress are the order in

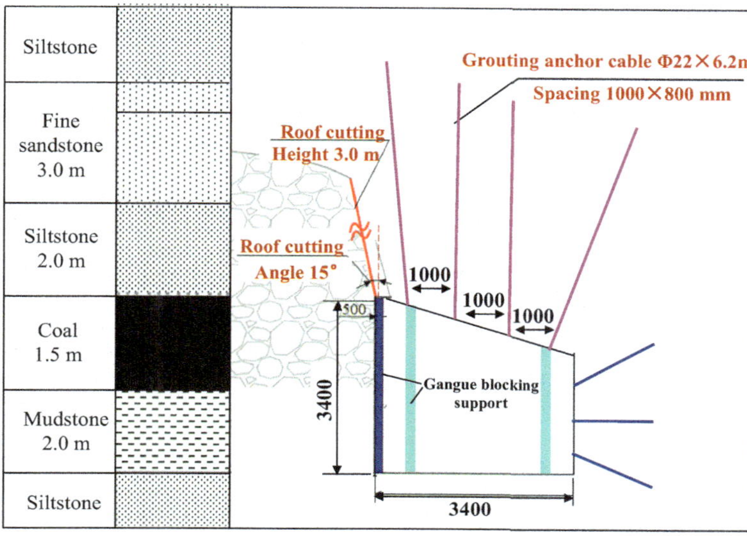

(a) Design diagram of roadway roof cutting and support parameters

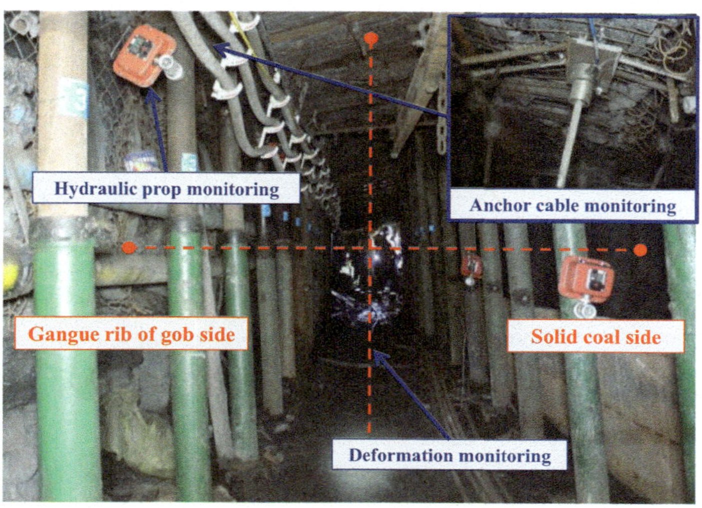

(b) Design scheme on-site diagram

Fig. 6.9 Design scheme of the AFR by roof cutting and bolt-grouting in Suncun Mine

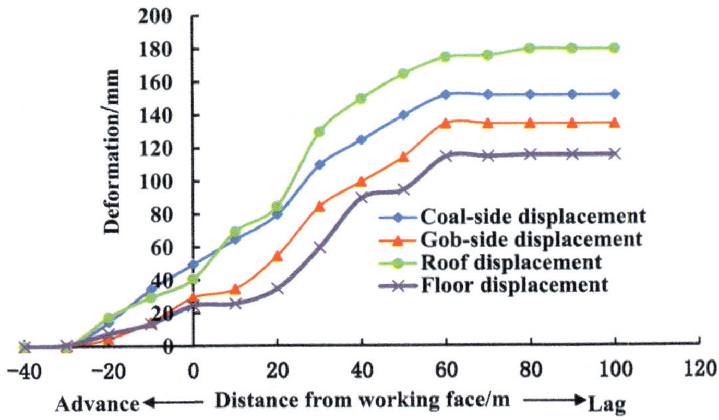

Fig. 6.10 Monitoring of the roadway deformation. Reprinted from Ref. Wang et al. (2020a, b), Copyright © 2020 Elsevier Ltd.

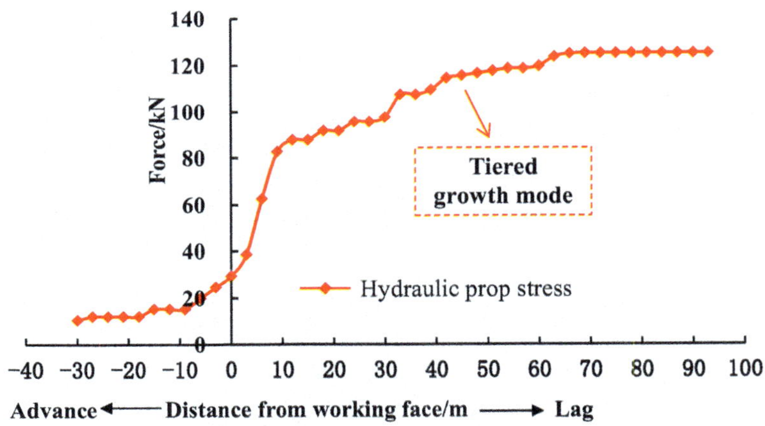

Fig. 6.11 Changes in the hydraulic prop stress. Reprinted from Ref. Wang et al. (2020a, b), Copyright © 2020 Elsevier Ltd.

which their magnitudes decreases. With a strength utilization ratio of 58.0% and a maximum stress of 238.5 kN, the anchor cable ultimately breaks at 411.0 kN, providing a sufficient safety margin.

The above monitoring data indicates that the AFR mining method effectively controls the safety and stability of deep roadway broken surrounding rock.

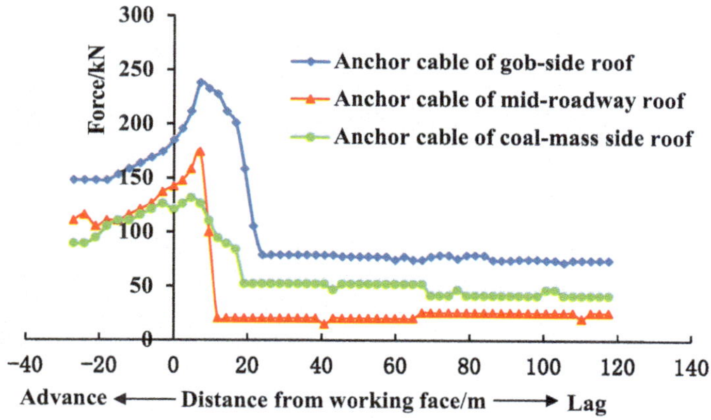

Fig. 6.12 Changes in the grouting cable stress. Reprinted from Ref. Wang et al. (2020a, b), Copyright © 2020 Elsevier Ltd.

Fig. 6.13 Field application effect of the AFR in Suncun Mine

6.5 The AFR with Deep Buried Depth

6.5.1 Engineering Object

The Guotun Mine, located in eastern China, is an example of a deep high stress mine with a maximum horizontal stress of 56 MPa. This mine is initially used by the gob-side entry driving mining method. The gob-side roadway is in an unfavorable high-stress condition. Additionally, traditional support materials are not able to absorb enough energy to cause large deformations in the roadway surrounding rock, which results in the failure of the support materials and significant resource waste from mine pillars. An example of a large deformation of a gob-side roadway driving resulting from the construction of a 5 m mine pillar at the Mine (buried depth: 890 m) is shown in Fig. 6.14.

Fig. 6.14 Surrounding rock large deformation in Guotun Mine. Reprinted from Ref. Xue et al. (2023), Copyright © 2023, Springer-Verlag GmbH Germany, part of Springer Nature

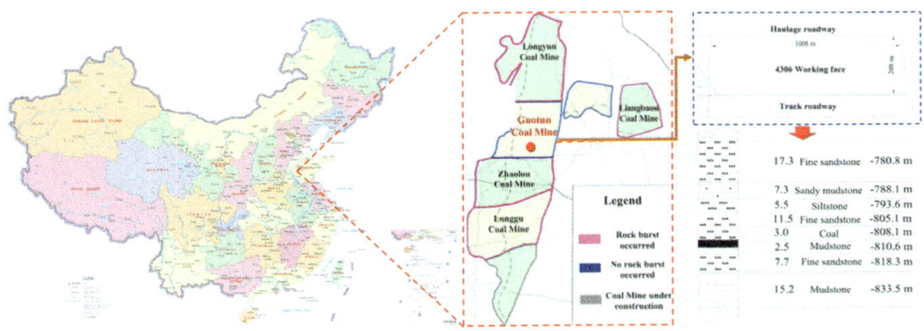

Fig. 6.15 General situation of Guotun Mine

The 4306 working face of the mine uses the AFR mining method in an attempt to address these issues. Coal seam has an average thickness of 2.95 m. This working face has a strike length of 1008 m and a dip length of 208 m. The immediate roof of the working face is 11.5 m fine sandstone, the basic roof is 5.5 m siltstone, and the immediate floor is 2.5 m mudstone. Figure 6.15 displays the geological and rock conditions of Guotun Mine's test working face.

6.5.2 Scheme Design

The AFR mining method is used in the track roadway of 4306 working face. Figure 6.16 displays the precise design specifications for roof cutting and energy absorbing support. The roof cutting height is 8.5 m, the roof cutting angle is 15°. The energy absorbing anchor cable has three rows and a design length of 10.8 m. There

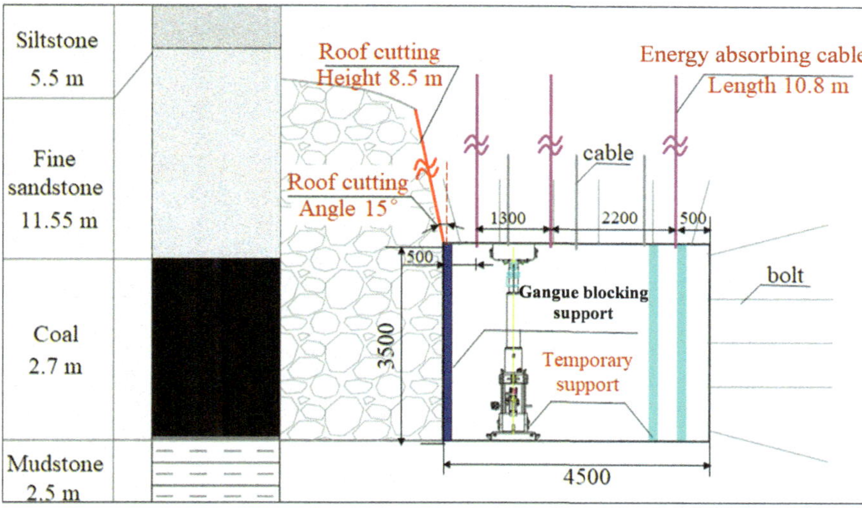

Fig. 6.16 Design diagram of roadway roof cutting and support parameters in Guotun Mine. Reprinted from Ref. Xue et al. (2023), Copyright © 2023, Springer-Verlag GmbH Germany, part of Springer Nature

are 800, 2400, and 4000 mm between rows, accordingly. The expandable U-shaped steel and steel mesh support the gob-side gangue rib, and the U-shaped steel has a support spacing of 500 mm.

6.5.3 Application and Monitoring Analysis

(1) Monitoring design

In order to verify the field application of the AFR mining method, a comprehensive monitoring section is installed to measure the energy, stress, and deformation of the surrounding rock of the working face and roadway as shown in Fig. 6.17. The followings are specific contents of the monitoring. A single hydraulic prop and hydraulic support are used for gangue rib support in the AFR. The specific monitoring layout parameters are as follows.

(i) Four micro seismic sensors spaced 200 m apart are placed in front of the working face to measure the energy levels of the surrounding rock. (ii) With regard to the coal body's stress monitoring, the borehole stress sensors are spaced 20 m apart on roadway mining coal side, with 8 m depth sunk. (iii) In terms of the working face's pressure monitoring, there are a total of 20 working face support stress sensors positioned 6 m apart on the working face. (iv) With regard to the support material's stress monitoring feature, the energy absorbing anchor cable has infrared anchor cable sensors attached at the end, with a 20 m monitoring section spacing. (v) In

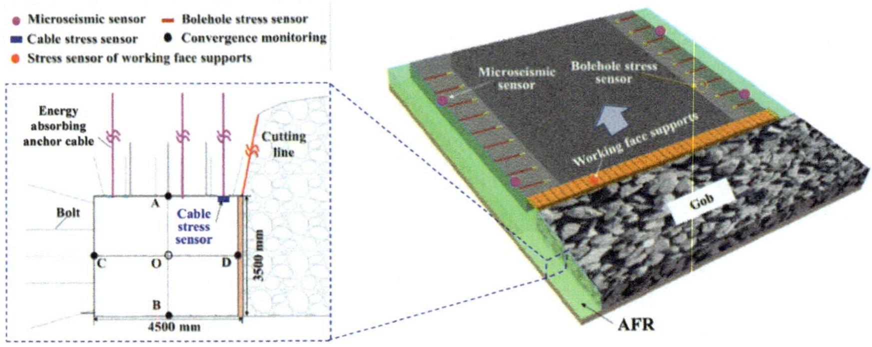

Fig. 6.17 Monitoring equipment arrangement. Reprinted from Ref. Xue et al. (2023), Copyright © 2023, Springer-Verlag GmbH Germany, part of Springer Nature

terms of the roadway's convergence monitoring, the monitoring section spacing is 20 m, which includes the intersection of the roadway's two sides and roof.

(2) **On-site mine pressure monitoring analysis**

The partition of the working face determines which two hydraulic supports are chosen for mine pressure monitoring. One support (#1) is situated in the region without the AFR mining method, while the other support (#120) is positioned in the region using the AFR mining method. Figures 6.18 and 6.19, respectively, display the load curves for #1 support and #120 support.

The monitoring of Figs. 6.18 and 6.19 shows that, in comparison to the area where the AFR mining method is not applied. The maximum pressure of the support in the area where it is applied is reduced by 3.3 MPa, or 6.7%. The average pressure of the support is reduced by 7.4 MPa, or 22.7%.

(1) **The Effects of Pressure Releasing**

Figure 6.20 displays the advancing abutment stress characteristic. The S3 stress sensor monitors the influence zone of roof cutting at 10.46 MPa, and its maximum

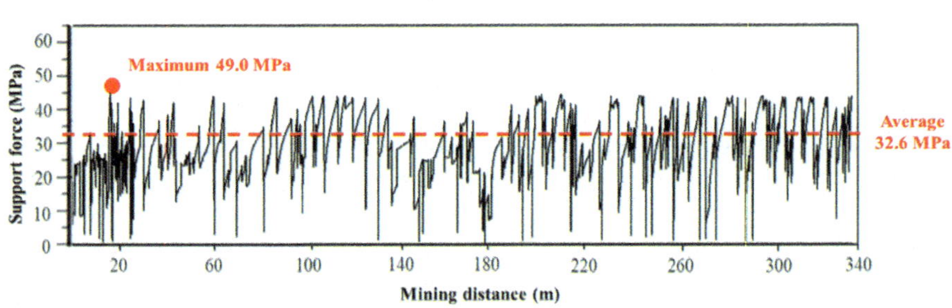

Fig. 6.18 Hydraulic support #1 force curve. Reprinted from Ref. Xue et al. (2023), Copyright © 2023, Springer-Verlag GmbH Germany, part of Springer Nature

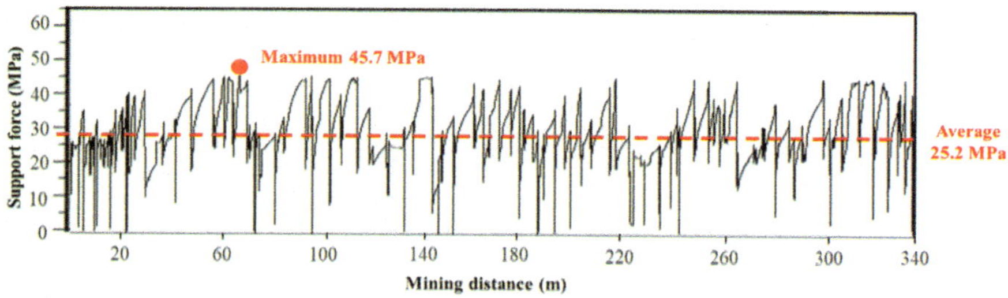

Fig. 6.19 Hydraulic support #120 force curve. Reprinted from Ref. Xue et al. (2023), Copyright © 2023, Springer-Verlag GmbH Germany, part of Springer Nature

stress is 16.06% smaller than that of the S33 stress sensor, which measures the influence zone of without roof cutting. Figure 6.21 displays the lateral abutment stress characteristic. Due to directionally roof cutting, the maximum stress in S4 (the influence zone of roof cutting) is 8.62 MPa, which is 15.46% less than in S34 (the influence zone without roof cutting). The advantages of directional roof cutting pressure relief in controlling surrounding rock are reflected in both aspects advancing abutment stress and lateral abutment stress.

(2) **Energy absorbing support characteristic**

The supporting force of a typical energy absorbing cable under large mine pressure is shown in Fig. 6.22. Figure 6.23 displays the energy absorbing and deformation data

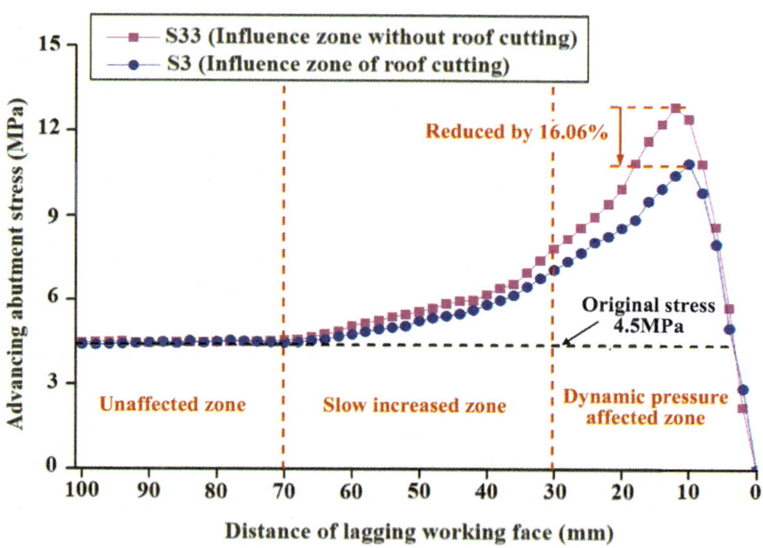

Fig. 6.20 Advancing abutment stress law. Reprinted from Ref. Xue et al. (2023), Copyright © 2023, Springer-Verlag GmbH Germany, part of Springer Nature

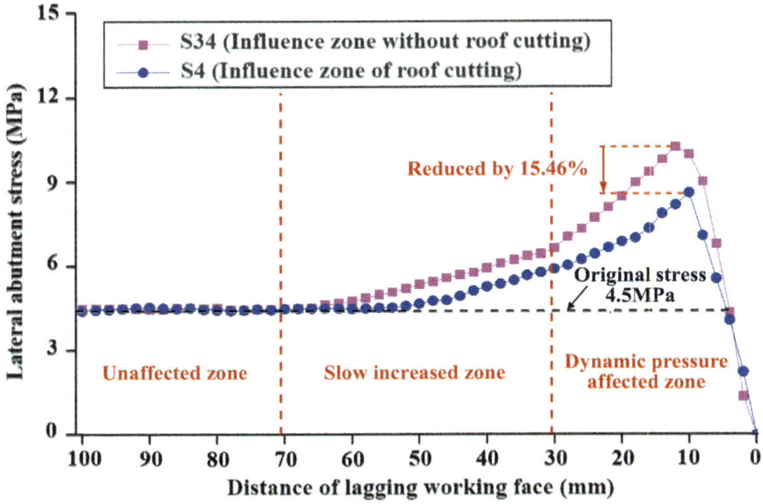

Fig. 6.21 Lateral abutment stress law. Reprinted from Ref. Xue et al. (2023), Copyright © 2023, Springer-Verlag GmbH Germany, part of Springer Nature

of six typical energy absorbing cables. The greatest elongation measured is 116 mm, and the associated absorbing energy is 4.06×10^4 J. The absorption energy is 2.52×10^4 J, and the average elongation is 72 mm. According to the study above, the energy absorbing cable has an exceptional ability to absorb energy and a strong capacity for rock mass deformation.

(3) **Analysis of surrounding rock displacement**

The displacement curve of the floor, roof, and two sides, which may be split into three stages, is displayed in Fig. 6.24.

There are three stages of deformation: (1) The first, known as rapid deformation, lags behind the working face by 0–150 m. (2) The second, known as slow deformation, lags behind the working face by 150–300 m. (3) The third, known as stable deformation, lags behind the working face by around 300 m. The study mentioned above may lead one to the conclusion that a stable region is reached 300 m after roof cutting, and at that time temporary support equipment may be taken down. The application effects of the AFR mining method are shown in Fig. 6.25. The maximum of roof and floor convergence deformation reaches 338 mm and the maximum of two sides of convergence deformation reaches 436 mm, which meets the safety production needs of the AFR.

The above monitoring data indicates that the AFR mining method effectively controls the safety and stability of roadway with deep buried depth.

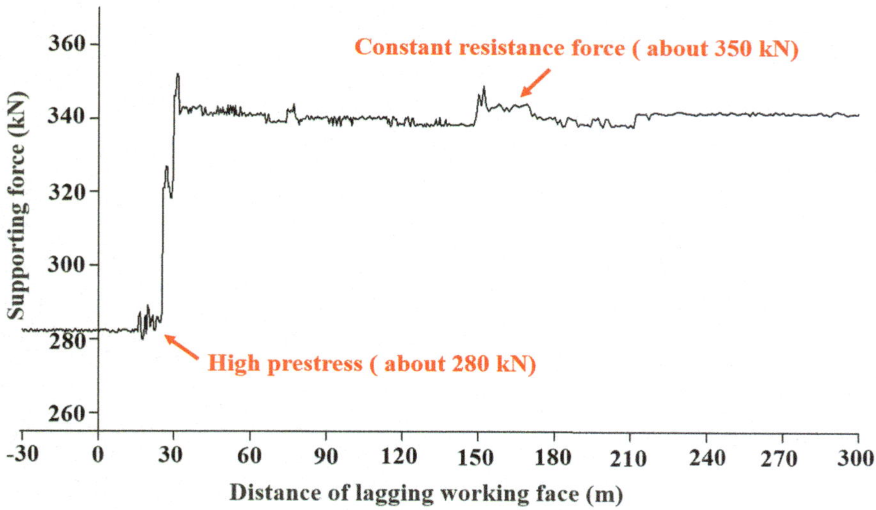

Fig. 6.22 Supporting force of energy absorbing cable. Reprinted from Ref. Xue et al. (2023), Copyright © 2023, Springer-Verlag GmbH Germany, part of Springer Nature

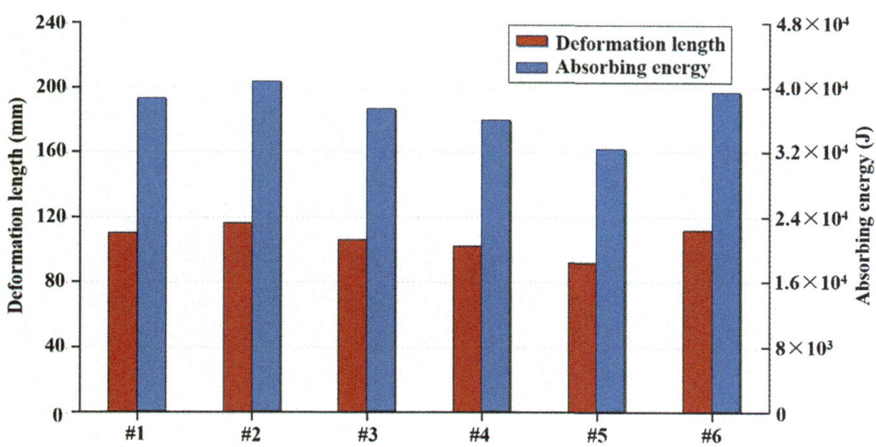

Fig. 6.23 The deformation and absorbing energy data. Reprinted from Ref. Xue et al. (2023), Copyright © 2023, Springer-Verlag GmbH Germany, part of Springer Nature

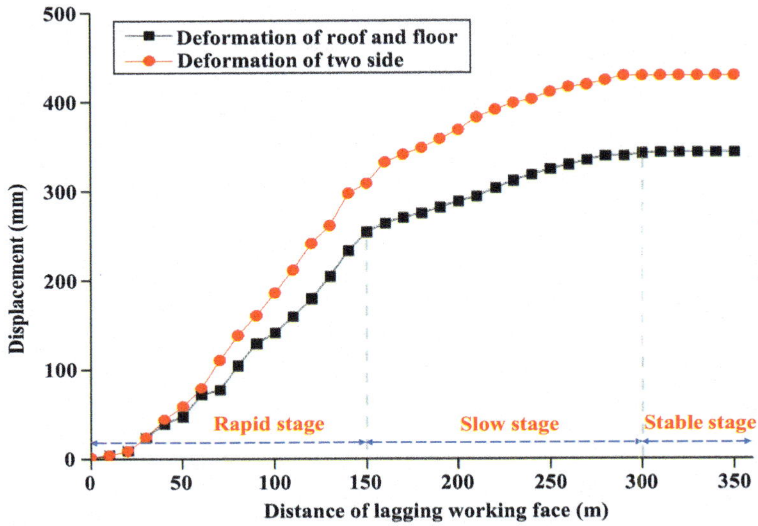

Fig. 6.24 Displacement at a monitoring station. Reprinted from Ref. Xue et al. (2023), Copyright © 2023, Springer-Verlag GmbH Germany, part of Springer Nature

(a) Formation stage of the AFR (b) Reuse stage of the AFR

Fig. 6.25 The application effects of the AFR mining method in Guotun Mine

6.6 The AFR Near the Overlying Coal Seam

6.6.1 Engineering Object

The AFR mining method has been successfully applied in Lujiayao Mine, near the overlying coal seam coal mine in western China. The 84,206 working face is located east of the 84,204 working face, with a mine pillar of 20 m reserved. The average buried depth of the working face is 270 m, the length of the working face is 223 m,

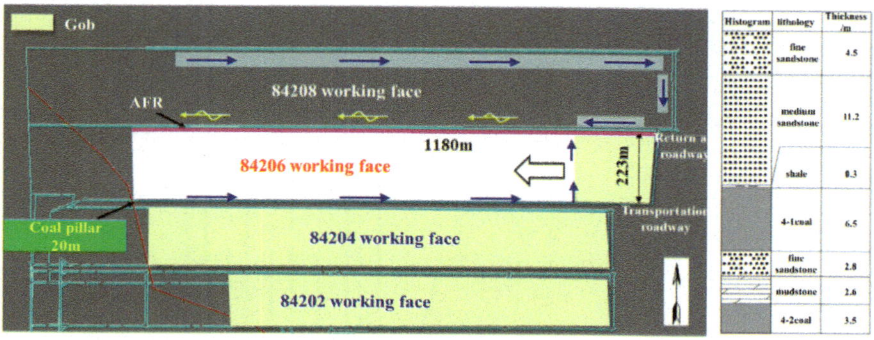

Fig. 6.26 Layout scheme of 84,206 working face in Lujiayao Mine

and the length of the AFR is 1180 m. The dip angle of the coal seam in the working face is 2°–5°, with an average of 3° and an average thickness of 3.0 m. The immediate roof is 2.6 m thick mudstone, and the basic roof is 2.8 m thick fine sandstone. The coal seam thickness is 6.5 m, a mining depth is 3 m, and a coal caving depth is 3.5 m. Therefore, the return air roadway of 84,206 working face, which applies the AFR mining method, is a typical near the overlying coal seam roadway, as shown in Fig. 6.26. The 5.4 m above the coal seam to be mined in this working face is the already mined coal seam, therefore, this working face is a typical near the overlying coal seam working face.

6.6.2 Scheme Design

Based on the engineering geological conditions of the working face and the construction requirements of the roadway surrounding rock, the following design parameters are finally determined (Fig. 6.27). (1) The roof cutting height is designed to be 5.0 m, with an angle of 15° from the vertical direction. (2) The roof is supported by energy absorbing cable, with a length of 4.7 m. (3) The gangue rib support by confined concrete is applied with a spacing of 0.5 m between confined concrete columns.

6.6.3 Application and Monitoring Analysis

The monitoring system for the AFR surrounding rock is shown in Fig. 6.28. The specific content includes the following. The monitoring systems lead by 100 m and lag by 300 m at the working face. Among them, three monitoring sections are arranged within a range of 100 m ahead, and three monitoring sections are arranged within a range of 300 m behind.

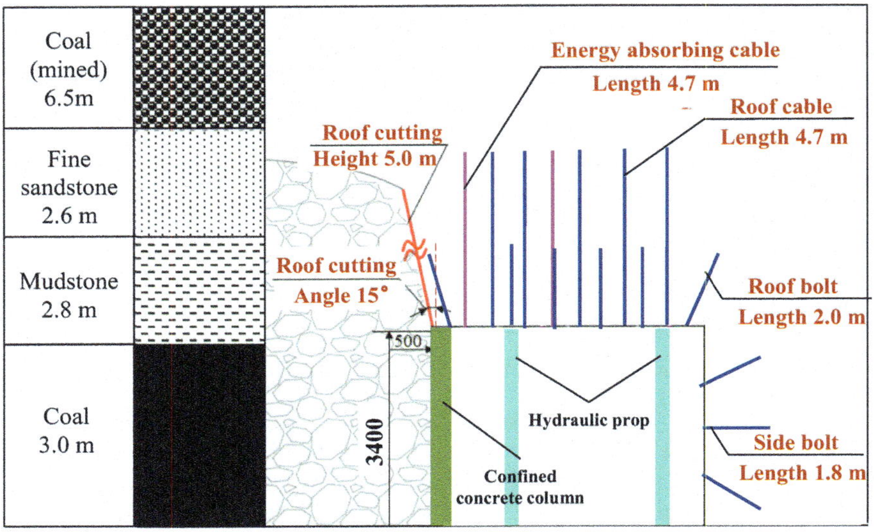

Fig. 6.27 Design diagram of roadway roof cutting and support parameters in Lujiayao Mine

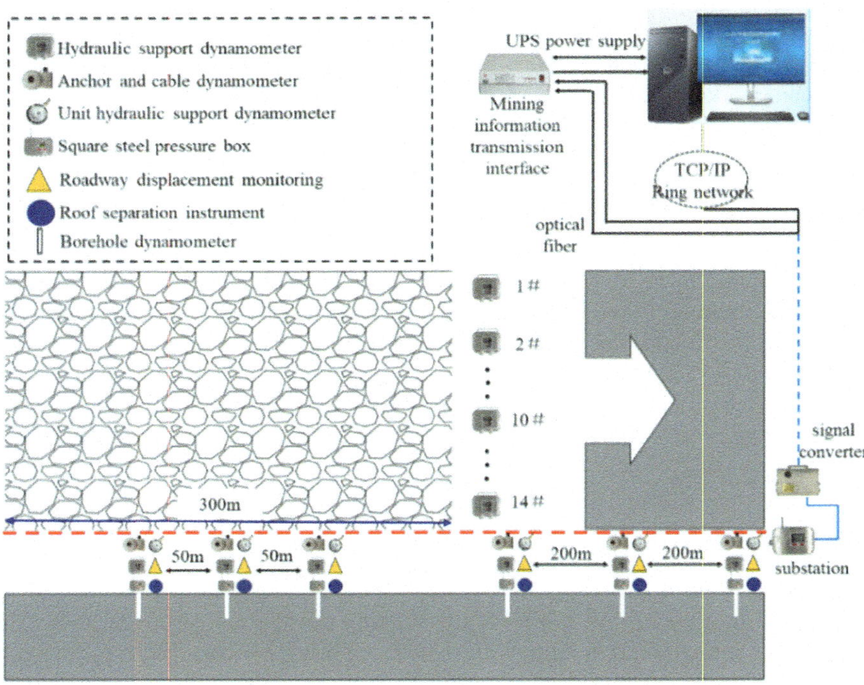

Fig. 6.28 Monitoring system for the AFR mining technology

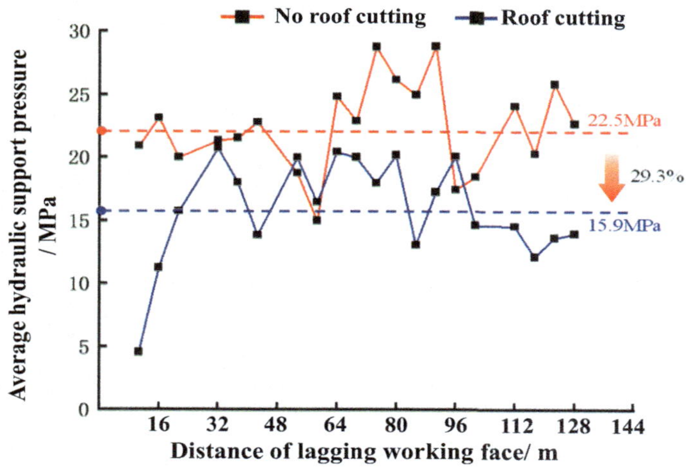

Fig. 6.29 Hydraulic support pressure of 84,206 working face

(1) Hydraulic support pressure

According to the monitoring results of mine pressure, there is a difference in the average support pressure between the non-roof cutting side (no using AFR) and the roof cutting side (using AFR). The hydraulic support pressure of 84,206 working face is shown in Fig. 6.29. The average roof pressure on the roof cutting side is 15.9 MPa, and The average roof pressure on the non-roof cutting side is 22.5 MPa. The roof pressure on the roof cutting side of the working face is generally reduced, which is 29.3% lower than that on the non-roof cutting side.

(2) Convergence deformation of roadway surrounding rock

The curves of convergence deformation of roadways at typical monitoring points are as shown in Fig. 6.30. Figure 6.31 depicts the on-site application effect of the AFR near the overlying coal seam.

From Fig. 6.30, the following analysis may be obtained.

1) The maximum value of convergence deformation of roof and floor is 146 mm, and the maximum value of convergence deformation of two sides is 93 mm at monitoring point 1. When the position of the monitoring point lags behind the working face by more than 72 m, the convergence deformation trend of the roadway weakens.
2) The maximum value of convergence deformation of roof and floor is 310 mm, and the maximum value of convergence deformation of two sides is 183 mm at monitoring point 2. When the position of the monitoring point lags behind the working face by more than 80 m, the convergence deformation trend of the roadway weakens.

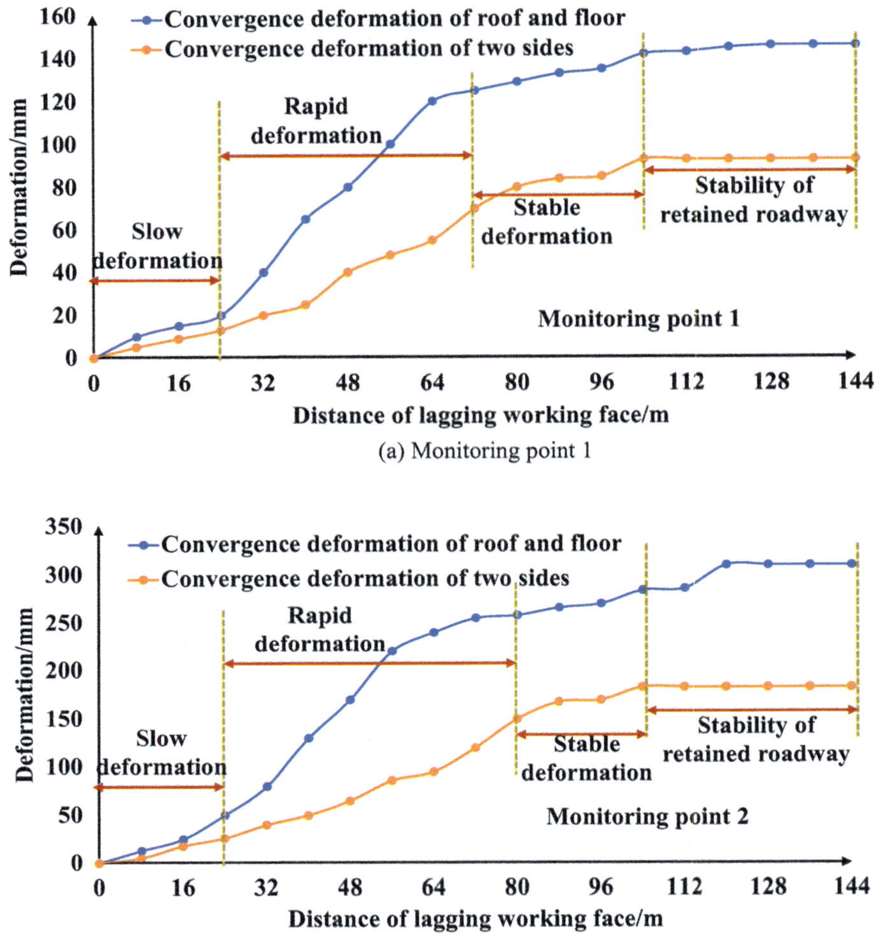

(a) Monitoring point 1

(b) Monitoring point 2

Fig. 6.30 Curve of convergence deformation of roadway

The above monitoring data indicates that the AFR mining method effectively controls the safety and stability of roadway near the overlying coal seam.

6.7 Chapter Summary

The successful application of AFR mining method in various complex conditions such as different buried depths, and roof types in mines is introduced. At present, it has been successfully applied in over 100 mines. This chapter focuses on different types

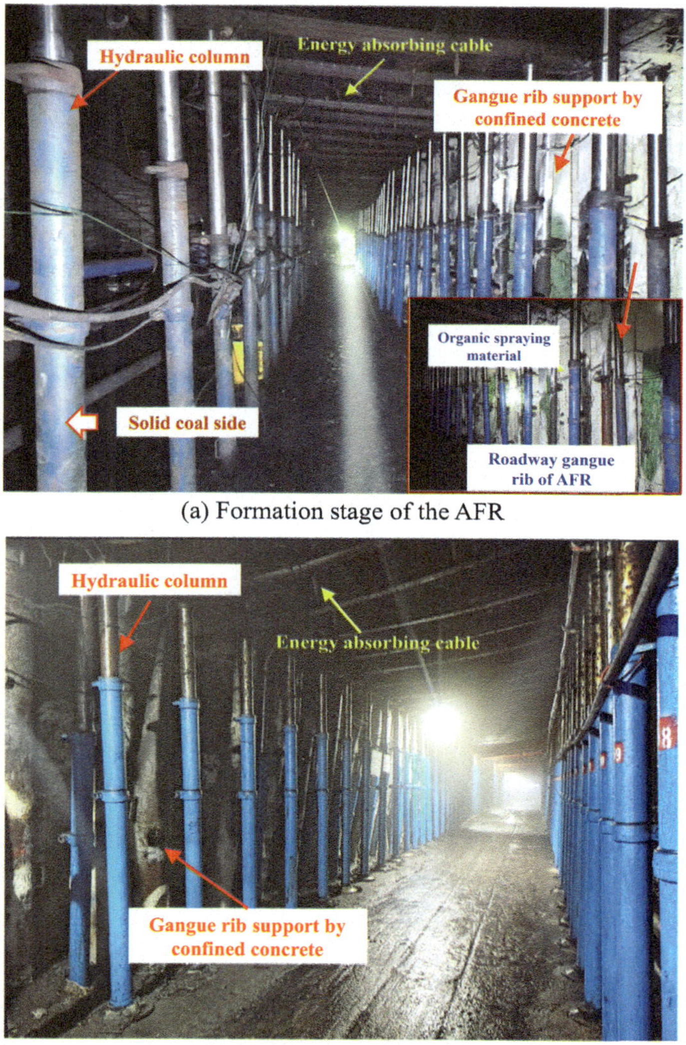

(a) Formation stage of the AFR

(b) Reuse stage of the AFR

Fig. 6.31 The application effects of the AFR mining method in Lujiayao Mine

of mine working faces including shallow buried depth, deep buried depth, weathered roof strata, broken roof strata, near the overlying mined coal seam and other typical examples. It effectively ensures the coal mining safety, eliminates mine pillars, and reduces roadway excavation, which has certain safety and economic advantages.

References

Wang Q, Jiang B, Xu S, He MC, Jiang ZH, Li SC, Wei HY, Xiao YC (2022) Roof-cutting and energy-absorbing method for dynamic disaster control in deep coal mine. Int J Rock Mech Min Sci 158:105186

Wang Q, Qin Q, Jiang B, Jiang ZH, He MC, Li SC, Wang Y (2020a) Geomechanics model test research on automatically formed roadway by roof cutting and pressure releasing. Int J Rock Mech Min Sci 135:104506

Wang Q, Jiang ZH, Jiang B, Gao HK, Huang YB, Zhang P (2020b) Research on an automatic roadway formation method in deep mining areas by roof cutting with high-strength bolt-grouting. Int J Rock Mech Min Sci 128:104264

Xue HJ, Wang Q, Jiang B, Jiang ZH, Wang Y, Huang YB, He MC (2023) Study on the method of pressure relief by roof cutting and absorbing energy in deep coal mine. Bull Eng Geol Env 82:198

Chapter 7
Summary and Prospect of AFR Mining Method

7.1 Application Scope and Hotspots

(1) The AFR mining method application scope

Through continuous research and exploration, the AFR mining method has achieved automatic roadway formation and safety control under different geological conditions. This method has been successfully applied in more than 100 mines, and engineering parameters of the applied mines is shown in Table 7.1.

In addition, the AFR mining method is applied in various complex conditions, such as different thicknesses of mine seams, buried depth of the mine, and different roof types in mines, and good application results have been achieved. Except for the engineering mentioned in Chap. 6, some typical application engineering images are shown in Fig. 7.1. This offers a productive and safe mining technique that may serve as a guiding theory and method for related mining engineering.

(2) The AFR mining method research hotspots

A search is conducted for pertinent publications on autonomously created AFR without mine pillars in the Web of Science Core Database. The change of research heat is reflected by the change trend of the number of papers published. The change trend of the number of papers published on the topic of AFR without mine pillars from 2011 to 2023 is shown in Fig. 7.2.

The first paper with the topic of "automatically formed roadway without mine pillars by roof cutting and pressure releasing" was published in 2013, and then the number of published papers showed a slow growth trend. Since 2017, the number of published papers has shown a rapid growth trend, indicating that the number of researchers studying this technology has begun to increase, and the research heat is constantly increasing.

The international journals that have published papers on th topic of AFR without mine pillars are mainly shown in Fig. 7.3. Among them, authoritative journals in

Q. Wang et al., *Theory and Method of Automatically Formed Roadway in Underground Mining*, https://doi.org/10.1007/978-981-97-8082-2_7

Table 7.1 Engineering parameters of the applied mines

Engineering parameters	Value
Buried depth of the mine	80–1200 m
Inclination of mine seam	0°–37°
Thickness of mine seam	0.92–4.1 m
Gas content in mine seam	0–20.4 m^3/t

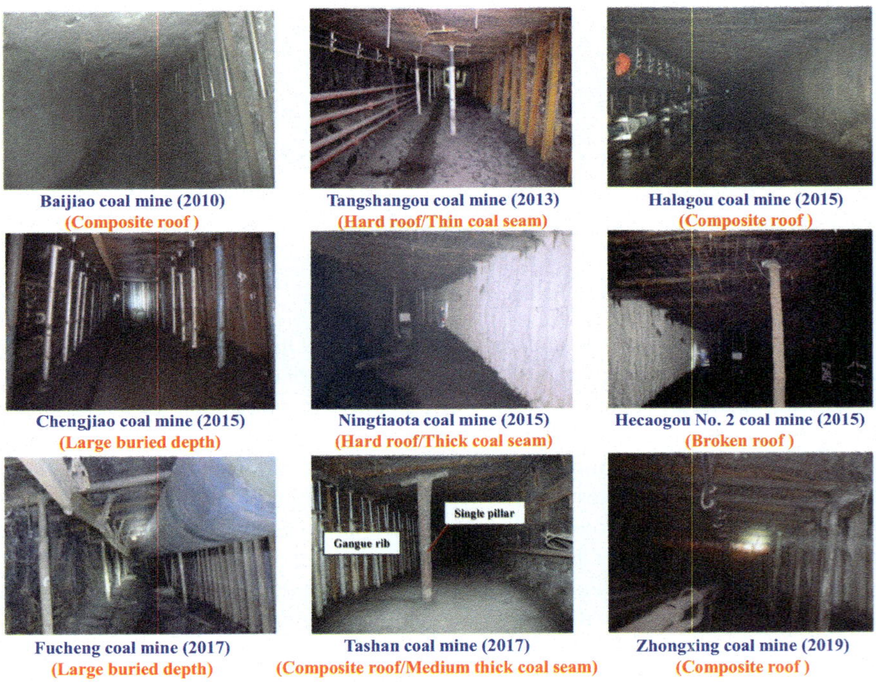

Fig. 7.1 Field applications in typical coal mines using the AFR mining method

the industry that publish papers on this topic mainly include International Journal of Rock Mechanics and Mining Sciences, Tunneling and Underground Space Technology, International Journal of Mining Sciences and Technology, and Journal of Rock Mechanics and Geotechnical Engineering, and so on. It shows that the papers on this topic have begun to be recognized by international journals, and the technology has begun to attract the attention of scholars in other countries.

In the WOS database, articles with a citation frequency in the top 1% of relevant academic fields are defined as highly cited articles, and articles with a citation frequency in the top 0.1% of relevant academic fields are defined as hot cited articles. Hot/Highly cited articles can reflect the innovation, recognition, and influence of research. In the WOS database, the hot/highly cited articles in the topic article on

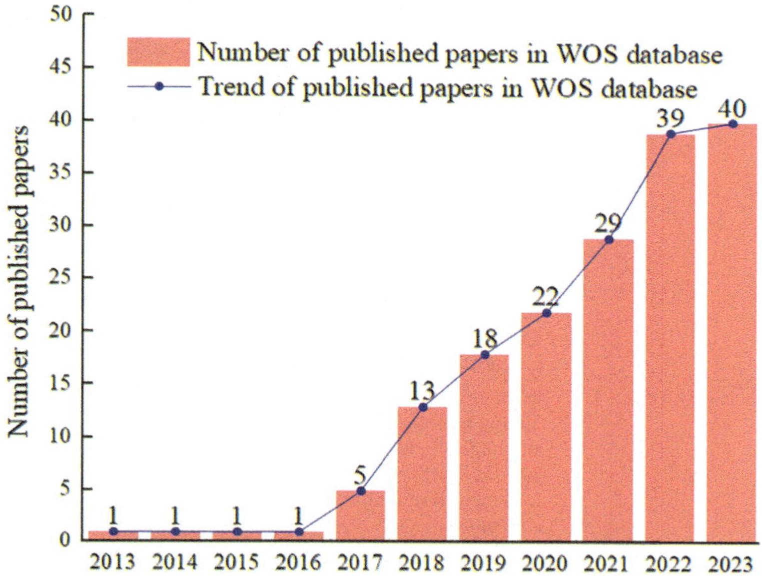

Fig. 7.2 Trends in the number of published papers

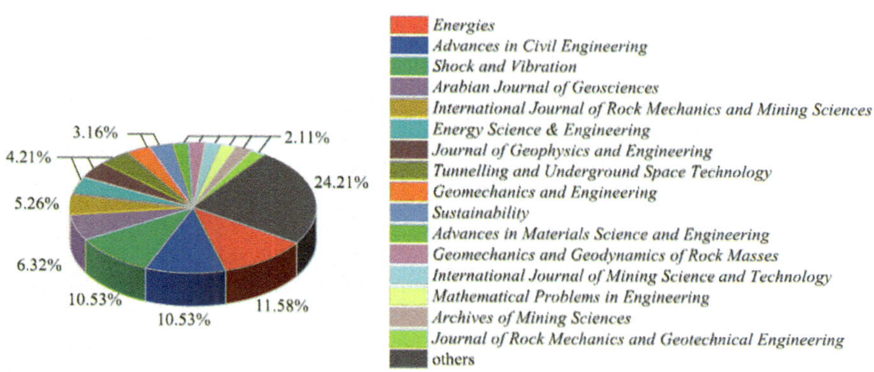

Fig. 7.3 Journals that have published papers on this topic

automatically formed roadway by roof cutting and pressure releasing are shown in Table 7.2.

The keywords in the article can reflect the research hotspots and development trend. Summarize the keyword changes in the articles on automatically formed roadway by roof cutting and pressure releasing in journals from EI database and from SCI database, as shown in Fig. 7.4.

From Fig. 7.4, it can be seen that from 2011 to 2014, the keywords in the article mainly focused on "Roof cutting pressure releasing technology", "Gob-side entry

Table 7.2 Hot/Highly cited articles in WOS database

Type	No.	Articles	Journals	First author
Hot cited articles	1	Comparative study of model tests on automatically formed roadway and gob-side entry driving in deep coal mines	International Journal of Mining Science and Technology	Wang Q
	2	Test research and application of automatically formed roadway without advance tunneling	Tunnelling and Underground Space Technology	Wang Q
Highly cited articles	3	Study of a no-pillar mining technique with automatically formed gob-side entry retaining for longwall mining in coal mines	International Journal of Rock Mechanics and Mining Sciences	Wang Q
	4	Research on an automatic roadway formation method in deep mining areas by roof cutting with high-strength bolt-grouting	International Journal of Rock Mechanics and Mining Sciences	Wang Q
	5	Innovation and future of mining rock mechanics	Journal of Rock Mechanics and Geotechnical Engineering	He MC

retaining technology", etc. There were fewer keywords in this stage, and research mainly focused on technology aspects. From 2015 to 2018, the keywords focused on "Short cantilever beam theory", "Roadway surrounding rock control", "Roof cutting key parameter", "Automatically formed roadway", and "Constant resistance and large deformation anchor bolt". At this stage, the number of keywords began to increase, and analysis of the roof cutting pressure relief mechanism was carried out. From 2019 to 2023, the key words mainly focused on "Balanced mining theory", "Roadway deformation mechanism", "Pressure releasing mechanism", "Coal pillar cancellation", "Deep mine pressure control", "Constant resistance support system", "Geomechanical model test", etc. At this stage, the balanced mining theory was officially proposed, and the number of keywords further increased. The breadth and depth of research continue to increase. The AFR mining method by roof cutting and pressure releasing entered a rapid development stage.

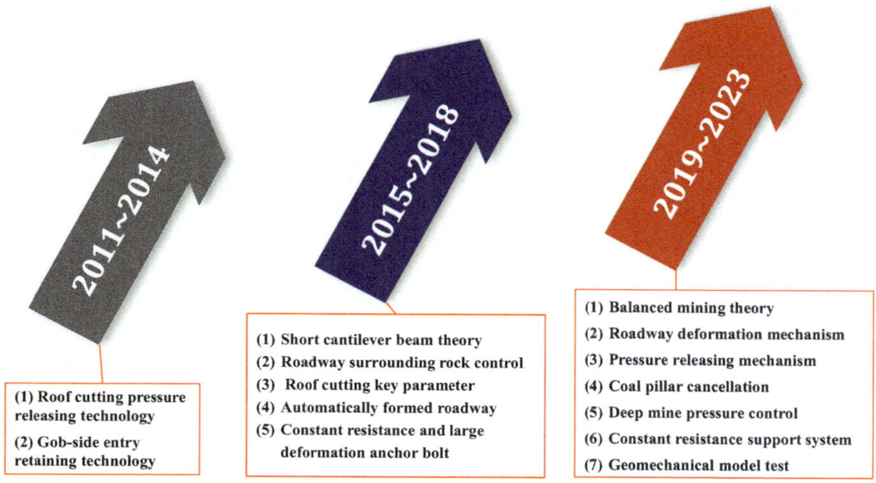

Fig. 7.4 Time curve of keywords

7.2 Future Research and Prospects

At present, the AFR mining method has a relatively complete theoretical, technical, and equipment system, and has been widely applied in the mining field, achieving good safety and economic benefits. The author provides the following prospects on the development of AFR mining methods in the future mining field. The following are specific prospects for (1) ultra-thick mine seam mining, (2) dynamic disaster prevention and control, (3) multi-dimensional roof cutting pressure releasing, (4) ecological protection, and (5) intelligent unmanned development.

(1) **The aspect of ultra-thick mine seam mining**

This method is relatively mature in the mining of thin and medium thick mine seam, and the mining of ultra-thick mine seam is still in the exploratory stage. Due to the need to ensure mining efficiency and the influence of the size of the coal mining machine, the working face of extremely thick coal seams needs to adopt a fully mechanized top coal caving mining technique, rather than a once mining full height technique. The fully mechanized top coal caving mining technique will leave a lot of coal in the gob, which is prone to spontaneous combustion, making it more difficult to seal the forming and reuse stages of the AFR.

Therefore, it is necessary to carry out in-depth research on the innovation of AFR without mine pillars in top coal caving mining for ultra-thick mine seam. Conduct research and development on the new technology of fully mechanized caving mining without mine pillars, optimize the ventilation system of the AFR, and develop new materials for gob sealing. The purpose is to achieve the application and promotion of the AFR mining technology without mine pillars in ultra-thick mine seams.

(2) **The aspect of dynamic disaster prevention and control**

In recent years, the depth of mining has been continuously increasing, and high stress and strong disturbance have led to an increase in the frequency of dynamic disasters such as rock burst and coal burst. The essence of dynamic disasters is the sudden release of energy. Through a series of studies on AFR mining method engineering in this book, it was found that this method has obvious advantages in reducing rock mass stress and energy. The specific advantages from this perspective are as follows. The AFR mining method solves the problem of stress concentration in mine pillars by eliminating the mine pillars at the root. This method also cuts off the high stress transmission path, leaving the AFR in a low stress environment. Furthermore, roadways are automatically formed by this method, avoiding the danger of roadway excavation in high stress areas. All of these can be seen, the AFR mining method provides new ideas for controlling dynamic disasters in deep mining processes.

Therefore, further research and development are needed on the key technologies for quantitatively reducing rock mass energy and regulating energy release by using AFR mining methods. In the future, the research on the prevention and control mechanism of the AFR mining method on the source of rock burst, mine coal burst and other disasters should be carried out in deep mines, so as to provide scientific theoretical guidance for the precise prevention and control of dynamic disasters in mines.

(3) **The aspect of multi-dimensional roof cutting pressure releasing**

The intensity of future resource mining will continue to increase, especially in multi seam mining, which will result in the concentration of stress in the surrounding rock being reflected in different heights of mine strata. As applying AFR mining methods, it is urgent to consider the pressure relief technology and methods for multiple mine strata.

Therefore, directional roof cutting and pressure relief technology with multiple spatial and temporal dimensions should be developed in the future. The spatial dimension includes the synchronous roof cutting and pressure relief of the upper and lower mine strata, while the temporal dimension includes the initial and secondary mining stages. It can optimize the stress and energy environment of the entire three-dimensional space of the mining area to the maximum extent.

(4) **The aspect of ecological protection**

Due to the mine pillars retention of different sizes in underground coal mining, the rock strata collapse to varying degrees, resulting in uneven subsidence of the surface and serious damage to the ecological environment. The AFR mining method proposed in this book is a mining method without mine pillar, providing ideas for solving the above issues.

Therefore, the mining pressure and rock broken expansion characteristics should be fully utilized to achieve the balance between the volume of collapsed gangue and mining volume in the future, effectively control the movement of roof strata, eliminate or reduce the surface damage, and achieve the balanced mining effect in AFR without mine pillars. In this way, efforts are made to minimize surface subsidence and protect the ecological environment as much as possible.

(5) **The aspect of intelligent unmanned development**

At present, in terms of working face mining, various equipment such as shearers, scraper conveyors, and hydraulic supports are used for linkage, indicating that the development of mechanization is already relatively complete. However, the process of roadway excavation is mostly manual construction, which limits the development of intelligent and unmanned mines. The AFR mining method proposed in this book is an automatically forming roadway method that eliminates roadway excavation, providing ideas for the more intelligent and unmanned development of mines. Meanwhile, the construction of transparent geological conditions is one of the important basic guarantees for precise mining and intelligent construction of mine resources.

Therefore, the intelligent equipment for roadway automatically forming and supporting of AFR mining method will be further upgraded and optimized to achieve ground control and automatic operation in the future. Meanwhile, intelligent comprehensive fine detection technology should be adopted to achieve intelligent mining transparency. A new mode of intelligent mining equipment layout should be constructed to achieve intelligent and unmanned mining of the whole mine.

The manufacturer's authorised representative in the EU is Springer Nature Customer Service Centre GmbH, Europaplatz 3, 69115 Heidelberg, Germany. If you have any concerns regarding our products, please contact ProductSafety@springernature.com

Printed and bound by CPI Group (UK) Ltd, Croydon, CR0 4YY
30/09/2026
02225308-0010